MOLECULAR BIOTECHNOLOGY

MOLECULAR BIOTECHNOLOGY

MUKESH PASUPULETI

Lecturer
Department of Biotechnology
Chaitanya degree and PG college
Hanamkonda, Warangal,
Andhra Pradesh, India

MJP PUBLISHERS
Chennai 600 005

Cataloguing-in-Publication Data

Pasupuleti, Mukesh (1978 –   ).
    Molecular Biotechnology / by
Mukesh Pasupuleti. – Chennai :
MJP Publishers, 2006.
    xx, 452p. ; 21 cm.
    Includes glossary and index.
    ISBN 81-8094-024-1 (pbk.)
    1. Biotechnology, Molecular I. Title
    660.6 dc22   PAS   MJP 017

ISBN 81-8094-024-1   **MJP PUBLISHERS**
Copyright © Publishers, 2006   A unit of Tamilnadu Book House
All rights reserved   47, Nallathambi Street
Printed and bound in India   Triplicane, Chennai 600 005

Publisher   :   J.C. Pillai
Managing Editor   :   C. Sajeesh Kumar
Project Coordinator   :   P. Parvath Radha

Edited and Typeset at [logo] Editorial Services, Chennai, eserve@rediffmail.com
Cover : R. Shankari   CIP : Prof K. Hariharan

To

My family members
*If not for you, the earth below
my feet will crumble and crack.*

PREFACE

Molecular biotechnology is now central to those various research fields grouped together under the heading "Molecular Sciences", and an ever-increasing number of new research students enter these fields every year. The primary objective of this book is that it should be accessible and informative to students encountering biotechnology for the first time.

Application of recombinant DNA technology to biology is bringing about a revolution in our understanding of living organisms. There is no field of experimental biology that is untouched by its power. We have just witnessed the publication of the draft of the Human Genome Project. Biotechnology has now become an integral part of every biology curriculum and some universities have begun to offer it as a specialized course.

This book is about the basic molecular mechanisms that underlie the principles of recombinant technology. It presents a concise coverage of the fundamental aspects of biotechnology in a readily understandable manner. I hope to capture the sense of understanding and anticipation that has followed the recombinant DNA technology breakthrough. A science that does not change is a dead science. Like all other authors I also faced the continuing task of updating previously written sections in order to have the final manuscript reasonably current. A book that lacks such updation will surely be unsatisfactory. I hope to obtain a unified, coherent approach and style to avoid the risk of being overwhelmed by the explosive pace at which new information has emerged. I have focused on selected specific areas that have already been studied in some depth and that illustrate the progress that has been made.

The topics that are covered and the approach make this book suitable for undergraduate and postgraduate students of molecular biology, cell biology, biochemistry, genetics, and biotechnology courses. Some additional material has been included in the hope that this will be useful to students.

I wish to acknowledge and thank all the authors who have contributed to biotechnology through their books and from which basic knowledge and information has been acquired and provided in this book.

I also thank Mr. Sajeesh Kumar, MJP Publishers, for ensuring that the chapters became a book rather than languishing in my computer.

I cannot express my gratitude in words to my parents for their support, which they rendered all these years. I am indebted to my sisters Kranthi, Sony and Chinnimayi for their support to combat the frustration, and for helping me to find the time to organize the chapters in an impressive manner.

I would also like to thank the many people, especially, Azeem-Ul Hussan, and Dr. S. Bandhyopadya without whose help or influence this book would not have been possible.

I hope that this book would find the patronage of teachers and students of all relevant fields of study. I hope that they will frankly communicate their comments/suggestions to me at mukesh_p78@yahoo.com. These will surely help me improve the book and make it more useful to the readers.

Mukesh Pasupuleti

Contents

PART I   MOLECULAR BIOLOGY

PART II   GENETICS

PART III   GENETIC ENGINEERING

PART I

MOLECULAR BIOLOGY

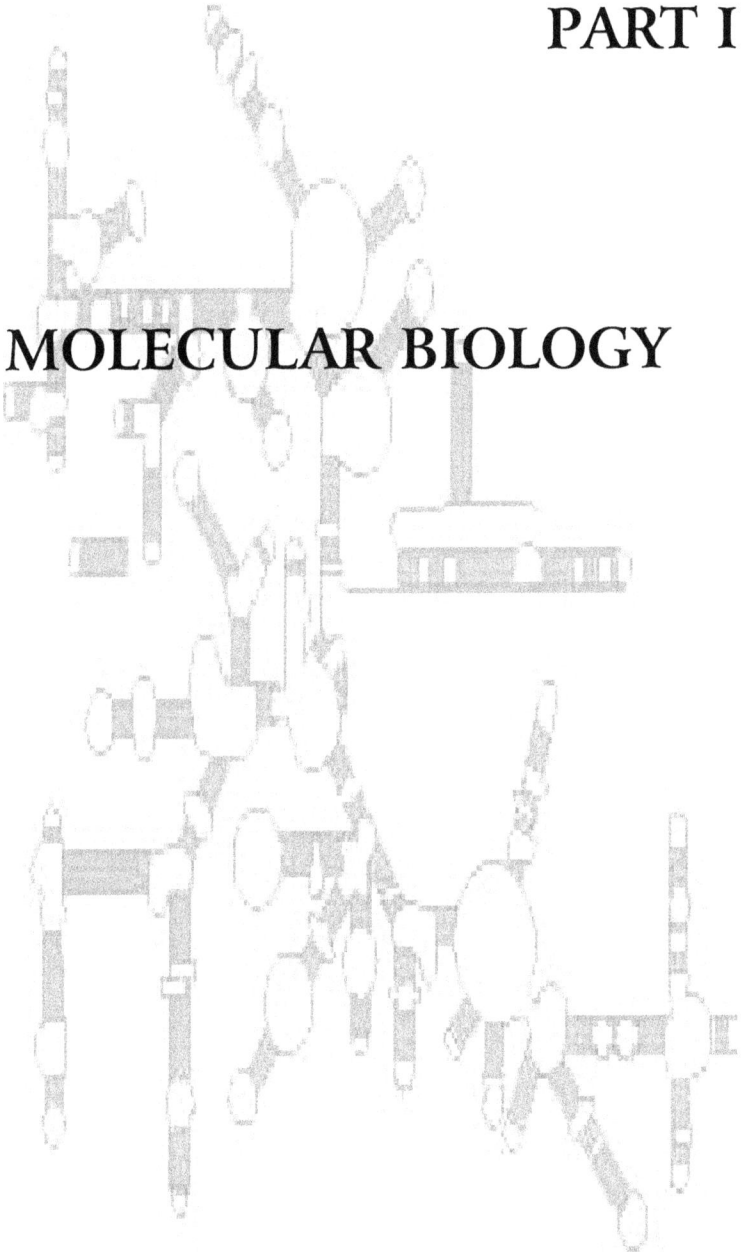

1

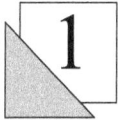

AN INTRODUCTION TO MOLECULAR BIOTECHNOLOGY

Molecular biotechnology is an exciting revolutionary scientific discipline that is based on the ability of a researcher to transfer specific units of genetic information from one organism to another. The objective of recombinant DNA technology is often to produce a useful product or a commercial process.

In early 1970, traditional biotechnology was not well known as a scientific discipline. Research in this area was carried out in chemical engineering departments. The term "biotechnology" was created in 1919 by a Hungarian engineer, Karl Ereky. According to Ereky, "biotechnology involves all works carried out with the aid of living things." More formally, biotechnology may be defined as "the application of scientific and engineering principles to the processing of material by biological agents to provide goods and services." In 1961, a Swedish microbiologist Carl Gordon Heden redefined biotechnology as "the industrial production of goods and services by processing using biological organisms, and it is firmly ground on expertise in microbiology, biochemistry and chemical engineering." However the nature of biotechnology was changed forever by the development of recombinant DNA technology. With these techniques, the optimization of any biotechnological process was achieved more directly.

Genetic engineering provided a means to create, rather than merely isolate highly productive strains. Microorganisms and eukaryotic cells could be used as "biological factories" for the production of insulin, interferon, growth hormone, viral antigens and a variety of other proteins. Recombinant DNA technology could also be used to facilitate the biological production of large amounts of useful lower-molecular-weight compounds and macromolecules that occur naturally in minute quantities. This technology facilitated the development of radically new medical therapies and diagnostic systems. Thus, the union of recombinant DNA technology with biotechnology created a vibrant, highly competitive field of study that has been called "Molecular Biotechnology."

Molecular biotechnology ought to contribute unprecedented benefits to humanity.

1. It should provide opportunities to accurately diagnose and prevent or cure a wide range of infectious and genetic diseases.

2. Significant increase in crop yield may be obtained by generating disease-, pathogen- or herbicide-resistant varieties.

3. Microorganisms that will produce chemicals, antibiotics, polymers, amino acids, enzymes, etc. can be developed.

4. Livestock and other animals that have enhanced genetically determined attributes can be developed.

Molecular biotechnology with much fuss and fanfare has become a comprehensive scientific venture, both commercially and academically, in a remarkably short time. A number of new scientific and business publications are devoted to molecular biotechnology. Both graduate and undergraduate programmes and courses have been created at many universities throughout the world.

Although it is exciting and important to emphasize the positive aspects of new advances, there are also social concerns and consequences that must be addressed. Because molecular biotechnology is so broadly based, its potential impact on society must be considered. Many issues have been considered and discussed extensively by government commissions, and debated by individuals in popular and academic publications. There has been an active and extensive participation by both scientists and the general public in deciding how molecular biotechnology should proceed, although some controversies still remain.

Finally, molecular biotechnology is the third scientific revolution after the industrial and computer revolution, which is going to change the lives and future of humankind on this earth. The ability to manipulate genetic material to achieve specified outcome in living organisms promises major changes in many aspects of modern life.

APPLICATION OF RECOMBINANT TECHNOLOGY

Genetics is the fundamental biological science and as every information lies in genes, there is no life without genes. A full understanding of any biological process can be achieved only when there has been a detailed analysis of gene structure and function. In olden days, for analysis of genes, researchers depended upon the mutations, which were time-consuming and cumbersome without any reproducibility. These problems have been solved by the advent of gene manipulation technologies, which provided novel solutions to experimental problems in biology. Now rDNA technology has risen to a level where its impact and presence is felt in every area of biology. Some of these applications are described in this section.

APPLICATION IN MEDICINE

Insulin Production

The first licensed drug produced through genetic engineering was human insulin an important hormone that regulates sugar metabolism. Insulin is produced by a small number of cells in the pancreas and secreted into the bloodstream. The inability to produce insulin results in diabetes, but daily injections of insulin are sufficient to reverse or at least avoid the effects.

In mammals, insulin is expressed as a single-chain prepro-hormone, which is secreted through the plasma membrane. A prepro-hormone contains extra amino acids not present in the mature hormone. Amino-terminal amino acids are present before the pre-sequence and target the expressed protein for secretion. The pro-sequence is a stretch of amino acids present before the pre-sequence and targets the expressed protein for secretion. The pre-sequence present in the middle of the chain is important for the current structure and function. During secretion, these extra amino acids are cleaved from the pre-hormone by cellular proteases to release the mature insulin molecule consisting of two short polypeptide chains, A and B, linked by two disulphide bonds. The primary problem in the production was getting insulin assembled into this mature form. This problem was solved by creating single-glycosidase-insulin fusion protein, which could be cleaved in a single step to release mature insulin.

Human Growth Hormone (hGH) Production

Children with hypopituitary dwarfism do not grow in size in normal proportions like children with normal pituitary function because their bodies make too little growth hormone. They are destined to be dwarfs unless they are treated with hGH. Growth hormone from other animals may not help. Moreover, early preparations of hGH were contaminated with pyrogens or other contaminants.

Growth hormone is a 191 amino acid protein that is produced in the pituitary gland and regulates growth and development. The production of hGH was achieved by constructing a hybrid gene from the natural hGH cDNA and synthetic oligonucleotides that encode the amino terminus of the mature form of the protein. This coding sequence was ligated into a plasmid adjacent to a bacterial promoter. Human growth hormone is produced by the bacteria and then secreted with the concomitant removal of the signal peptide by bacterial proteases. The only difference between the secreted hGH and that produced intracellularly is the presence of an amino-terminal methionine on the intracellular expressed molecule. Because the secreted form lacks this methionine, it is called met-less hGH.

Improved Vaccines

Novel, safer vaccines may be another product of cloned genes. Usually, vaccines are made from whole viruses that have been either killed or weakened so that they can no longer cause disease but can still stimulate the body's immune defences. The problem is that sometimes these viruses are not killed or are not weakened enough, thus they end up causing the disease and sometimes even death.

The first successful subunit vaccine was produced for hepatitis B virus (HBV), which infects the liver. Initial attempts to produce the HBsAg protein in *E. coli* failed, so researchers turned to yeast. The HBsAg gene was inserted into a high-copy yeast expression vector. By growing the yeast in large fermentors, it was possible to produce 50–100 mg/l of protein closely resembling the natural viral protein. The yeast protein is now used commercially to vaccinate people against HBV infection.

The most successful vaccination campaign to date using a recombinant vaccine is the elimination of rabies with a vaccinia-based vaccine. This has been administered to the wild population of foxes in a large part of central Europe by giving food consisting of chicken heads spiced with the recombinant virus. The epidemiological effect of vaccination has been most evident in eastern Switzerland where there was a drastic decrease in the number of deaths due to rabies virus.

Complex Human Proteins

Most of the proteins are simple and can be expressed in bacteria and yeast. But proteins of medical interest are considerably more complex in structure, and it is difficult to produce biologically active proteins in bacteria and yeast. Mammalian cells can be useful but they are ficky and expensive to grow. Thus much of the effort has been devoted to setting up fermentor systems for large-scale culture of mammalian cells.

The first drug to be produced commercially by mammalian cell culture was tissue plasminogen activator or tPA, which is administered to heart attack victims. Tissue plasminogen activator is a protease, an enzyme that cleaves other proteins. Rapid administration of a plasminogen activator after a heart attack dissolves the life-threatening clots that lead to irreversible damage of heart muscle. Tissue plasminogen activator is commercially produced from a mammalian cell line carrying a stable, integrated, highly amplified expression vector. Another protein being produced by mammalian cell culture is factor VIII, a protein required for normal clotting of the blood. The factor VIII cDNA has already been cloned and is commercially available.

Identification of Disease-causing Microorganisms

A disease can be caused by a battery of microorganisms. In conventional diagnostic practice, identification of the causative organisms would require cultivation of samples on a variety of different media in a variety of different ways like microscopy, animal cell culture and immunoassay. These procedures are time-consuming,cumbersome and disparate. This whole set of disparate test procedures can be replaced with a single technology called hybridization. Clearly, hybridization of the test sample with a batteryof probes is more simple and less time-consuming. Hybridization techniques can be used when the infectious agent is present in very low concentration and when the organism cannot be cultured. An important diagnostic tool is *in situ* hybridization for detecting viral pathogens present in the cell.

Hybridization as a diagnostic tool is not restricted to clinical microbiology. There are many applications in plant pathology, monitoring disease control methods and microbial ecology. The disadvantages of nucleic acid probes are that they can determine the presence of microorganisms but cannot tell whether it is dead or alive.

Detection of Genetic Disorders

There are several hundred recognized genetic diseases in man which result from single recessive mutations. In some cases, a protein product that is defective or absent has been identified but in many others the nature of the mutation is unknown. For many of these genetic diseases there is no definite treatment and their prevention is the current strategy in many countries.

An essential prerequisite for prenatal diagnosis is the availability of foetal DNA. Foetal cells can be obtained by amniocentesis but this method is not entirely satisfactory for it cannot be carried out early in pregnancy. An alternative approach is to obtain foetal DNA from biopsies of trophoblastic villi in the first trimester of pregnancy. DNA amplification is done without purification of the cellular DNA with PCR. Of all genetic diseases, the inherited haemoglobin disorder has been the most extensively studied at the DNA level. In what follows, haemoglobinopathies will be taken as examples. Their antenatal diagnosis by recombinant DNA techniques has served as a prototype for other genetic disorders. Clinically the most important haemoglobinopathies are sickle-cell anaemia and the thalassaemia site in genomic DNA. This can be used as a marker for the presence or absence of the defect. Digesting the mutated and normal DNA with the restriction enzyme and performing a Southern-blot hybridization with a cloned β-globulin DNA probe can therefore detect the mutation. Such an approach is applicable only to those disorders where there is an alteration in a restriction site or where a major deletion or rearrangement alters the restriction pattern.

Monoclonal Antibodies as Magic Bullets

Researchers have long dreamed of harnessing the specificity of antibodies for a variety of uses that require the targeting of drugs and other treatments to particular sites in the body. It is this use of antibodies as targeting devices that led to the concept of the "magic bullet", a treatment that could effectively seek and destroy tumour cells and infectious agents wherever they resided.

The major limitation in the therapeutic use of antibodies is producing a useful antibody in large quantities. Initially, researchers screened myelomas which are antibody-secreting tumours for the production of useful antibodies. But they lacked a means to programme a myeloma to produce an antibody to their specification. This situation changed dramatically with the development of monoclonal antibody technology. The procedure for producing monoclonal antibodies or mABs involves inoculation of the desired antigen into a BALB/C mice after the animal mounts an immune response to the antigen which occurs after 21 days. The animal is sacrificed and its spleen is removed. Spleen cells obtained by mashing the spleen are fused to myeloma cells. The resulting fused cells or hybridomas retain properties of both parents, grow continuously and rapidly like cancer cells with antibody-producing ability. Hundreds of hybridomas can be produced from a single fusion experiment and they are screened to identify clones which produce the antibody. Once identified, this antibody is available in limitless quantities.

Protein Engineering

One of the most exciting aspects of recombinant DNA technology is that it permits the design, development and isolation of proteins with improved operating characteristics and even completely novel proteins. The simplest example of protein engineering involves site-directed mutagenesis to alter key residues. The thermostability of T4 lysozyme was increased 100-fold by the introduction of a disulphide bond.

Many human proteins are being tested as potential therapeutic agents and a number of them are already commercially available. Protein engineering now is being used to generate second-generation variants with improved pharmacokinetics, structure, stability and bioavailability. Insulin is most likely to be assembled as zinc-containing hexamers. This self-association may limit absorption. By making single amino acid substitutions, insulin with high activity and faster absorption was made. Replacing asparagine residue with glutamine altered the glycosylation pattern of tissue plasminogen activator. This in turn significantly increased the circulatory half-life, which in the native enzyme is only 5 min.

APPLICATIONS IN AGRICULTURE

Recombinant DNA technology has not only enhanced the health of humans, but has also contributed to exciting developments in agricultural biotechnology. Using rDNA methods, transgenic plants and animals with desirable properties such as resistance to diseases/herbicides have been developed. Flowers with exotic shapes and colours have been genetically engineered by transgenic expression of pigment genes. Recombinant growth hormones are now available for farm animals, resulting in leaner meat, improved milk yield and more efficient feed utilization. In future, transgenic plants and animals may serve as bioreactors for the production of medicinal or protein pharmaceuticals.

Expression of Viral Coat Protein to Resist Infection

Viruses are a serious problem for many of the agricultural crops and animals. Infections can result in reduced growth, yield of products and quality. Through a standard genetic trick termed cross-protection, infection of a plant/animal with a strain of virus that produces only mild effect protects the plants/animals against infection by more damaging strains. The same principle/ mechanism when applied to animals is called as vaccination. Although the mechanism of cross-protection is not entirely known, it is thought that a particular viral-encoded protein is responsible for the protective effect.

Expression of Bacterial Toxin

Currently, the major weapons against the attackers of plants are chemical insecticides. But chemicals have some impact on the environment. Natural microbial pesticides, such as the species *Bacillus thuringiensis (Bt)* have been used in a limited way for over 38 years. Upon sporulation, these bacteria produce a crystallized protein that is toxic to the larvae of a number of insects. The toxic protein does not harm non-susceptible insects and has no effect on vertebrates. The crystal protein is normally expressed as a large, inactive pro-toxin about 1200 amino acids in length and with a molecular weight of 1,20,000 Dalton. The toxin acts by binding to receptors on the surface of mid gut cells and blocking the functioning of these cells. Research is on the way for commercial use.

A second approach to the development of insect-resistant plants has been through the transgenic expression of serine protease inhibitors. These proteins are present in a number of plants and function to deter insects by inhibiting serine proteases in the insect digestive system.

Herbicide-Tolerant Plants

The presence of weeds in a crop field reduces the yield. Weed killers or herbicides are not very selective and their current use relies on differential uptake between the weed and the crop plant or on application of the herbicide before planting a field, altering the food content of plants. With the ability to introduce DNA into plants, researchers are trying to create herbicide-tolerant crops by three strategies.

1. Increasing the level of the target enzyme for a particular herbicide.

2. Expressing a mutant enzyme that is not affected by the compound.

3. Expressing an enzyme that detoxifies the herbicide.

Of the large number of herbicides in use today, only a few of the cellular targets have been characterized. The first strategy is to clone the cellular target genes into the plant so that they are produced in large amounts.

Another strategy for creating herbicide-tolerant plants has used mutant forms of bacterial EPSPS enzymes. Genes encoding these mutant enzymes have been cloned from glycophosphate-resistant bacteria and expressed in plants. These enzymes have lesser inhibition effect by herbicides. The third strategy for creating herbicide-tolerant plants is by transgenic expression of enzymes that convert the herbicide to a form that is not toxic to the plant. Some plants have developed their own detoxifying system for certain herbicides. But these activities in plants are encoded by a complex set of genes that has not yet fully characterized.

Pigmentation in Transgenic Plants

Plants are widely used for ornamental purposes, so it is not surprising that considerable attempts have been made to develop varieties having flowers with new colour, shape and growth properties can be engineered. Pigmentation in flowers is mainly due to three classes of compounds, the flavonoids, the carotenoids and the betalains. Of these, the flavonoids are the best characterized with much information now available concerning their chemistry, biochemistry and molecular genetics.

Experiments are underway to expand the spectrum of colouring of certain floral species by introducing the genes for the entire pigment biosynthesis pathway. A blue rose has never been obtained because rose plants lack the enzymes that synthesize the pigment for blue flower colouration. But introducing the genes for blue colour gave very few successful results. This is due to a phenomenon called co-suppression, in which an extra copy of the gene will suppress the expression of the endogenous genes. An experiment was performed in which a second copy of a petunia pigment gene was introduced into a petunia plant with coloured flowers. It is expected that increased production of the encoded enzyme might produce flowers with a deeper purple colour. But white coloured flowers where produced due to co-suppression. Co-suppression now has been demonstrated in numerous other systems. It does not appear to be a dosage effect resulting from competition for transcription factors. Nor is it a result of a system that detects specific duplicate plant genes. Rather it appears to be the result of a homology-dependent interaction between homologous sequences.

Altering the Food Content of Plants

Starch is the major storage carbohydrate in higher plants. A wide range of different starches is used by the food and other industries. These are obtained by sourcing the starch from different plant varieties coupled with new enzymatic or chemical approaches to creating novel starches with new functional properties.

Higher plants produce over 200 kinds of fatty acids, some of which have value as food. However, many are likely to have industrial (non-good) uses of higher value than edible fatty acids. These are widely used in detergent synthesis. Phytate is the main storage form of phosphorus in many plant seeds but bound in this form, it is a poor nutrient for monogastric animals. Plants with phytase gene will produce seeds with lower phytate content and higher phosphorus content. Supplementation of broiler diets with transgenic seeds resulted in improved growth rate comparable to diets supplemented with phosphate or fungal phytase.

Transgenesis

Many generations of selective mating are required to improve livestock and other domesticated animals genetically for traits such as milk yield, wool characteristics and rate of weight gain. The selective mating procedure is time-consuming and costly. Until recently, selective breeding was the only way to enhance the genetic features of domesticated animals. The process of gene manipulation to permanently modify the germ cells of animals is called as "Transgenesis".

The animal is called as "transgenic" animal. However, transgenic mice have been produced which carry genetic lesions identical to those existing in certain human inherited diseases. Such mice can be used as models for the development and evaluation of new pharmaceutical entities. The power of this technique can be illustrated by its application to studies on tumour development. In addition to their use as animal models of human diseases, transgenic mice can be used for mutagenicity testing. For many years, mice have been used in long-term toxicity testing of new chemicals. In these tests, the animal is acutely or chronically exposed to the test compound and observed constantly for occurrence of tumours.

The most obvious uses of transgenic animals in food production, is when it produces novel products. Fortunately, significant progress has been made in producing pharmaceutical proteins in the milk of transgenic animal forms. Apart from the greater regulatory acceptability of farm animals, they offer the potential of high volumetric productivity since milk contains tens of grams of proteins like casein, β-lactoglobulin and whey acidic protein per litre. Production of these proteins is tightly regulated by promoters that limit genes to express only in cells of the mammary gland. The regulatory sequences from the genes of milk-specific proteins have been cloned along with the gene of interest and used to control the expressions of genes in transgenic animals.

One of the goals of transgenesis of dairy cattle is to change the constituents of milk. The amount of cheese produced from milk is directly proportional to the K-casein content. Increasing K-casein production with an over-pressed K-casein transgenic is a reasonable likelihood. For a different end use, heterologous expression of a lactose transgene in the mammary gland could result in milk that is free of lactose.

Transgenic animals may sometimes serve as bioreactors that continually secrete high levels of desired proteins into their milk. The proteins would be harvested simply by milking the animals and then standard chromatographic procedures would be used to purify it. The potential applications of transgenic animals in medicine have been expanded with the expression of proteins in blood, tissues and animal organs for transplantation to humans.

INDUSTRIAL APPLICATIONS

Recombinant DNA technology does not just offer novel methods for the generation of proteins, it also provides new ways of making low molecular weight compounds. A good example is the microbial synthesis of the blue dye indigo and the black pigment melanin. Neither of these compounds is normally produced by microbes. The cloning of a single gene from *Pseudomonas putida*, that encodes naphthalene dioxygenase, results in the generation of an *E. coli* strain able to synthesize indigo in a medium containing tryptophan. Just as novel proteins can be produced by recombinant DNA techniques so too can novel small molecules. The *Streptomyces coelicolor* gene cluster encoding the biosynthesis of the isochromane-quinone antibiotic was cloned into a variety of other *Streptomyces* spp. producing different isocheomane quinines. At least three new antibiotics were detected. Clearly, actinoradin, a novel metabolite present in the *Streptomyces* sp., is subjected to further or different modifications. Thus, a new class of antibiotics such as 2–norethyromycins A, B, C and D and isovaleyl spiramycin are synthesized.

Biotechnology is going to be the next scientific revolution after the industrial revolution, which is going to change the lives and future of humankind on this earth. The ability to manipulate genetic material to achieve specified outcomes in living organisms promises major changes in many aspects of modern life.

2

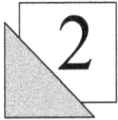

GENETIC MATERIAL

Today, if anyone is questioned as to what the genetic material is, there would be only a single answer—DNA. But the same question when asked during 1900, would have generated a greater variety of answers some saying it as RNA, some saying protein and some arguing DNA as hereditary material or genetic material. But everyone believed that the hereditary material is genes, but diverged on the chemical nature of genes.

Genes are the central concept of the heredity science, genetics. If genes are physical particles, like other cell components, they must be made of molecules and it should therefore be possible to study them directly by biophysical and biochemical methods. This led to a new branch of genetics called Molecular Biology. Genetics and molecular biology are very closely related subjects and although there is a distinction between them, there are more similarities. For this reason, the term molecular genetics is now often used to describe that branch of biology concerned with the study of all aspects of genes. The initial aim of molecular genetics was the identification of the chemical nature of the gene. This new approach led to new concepts, and soon biologists ceased to regard individual genes simply as units of inheritance and instead began to look on them as biological information with the entire complement. Genes in an organism contain the total amount of information needed to construct a living, functioning body of that organism. Genes themselves are the instructions for making other biomolecules. Hence to understand the structure and functions of genes as molecules, we must understand the qualifications or prerequisites a molecule must have to qualify as genetic material.

NATURE OF GENETIC MATERIAL

Freidrich Miescher isolated nuclei from pus cell in waste surgical bandages. He found that nuclei contained a novel phosphorus-bearing substance, which he named as nuclein. Nuclein is mostly chromatin, a complex of deoxyribonucleic acid (DNA) and chromosomal proteins. A range of experimental techniques mostly cytochemistry suggested that either DNA or protein or both must be the genetic material. The most fundamental property of the genetic material is that it must be able to exist in an almost infinite variety of forms: each cell contains a large number of different genes, each controlling a different heritable trait and each presumably having a structure slightly different from any of the other genes in the cell. Apart from infinite varieties, genetic material must be capable of precisely directing its own replication so that every daughter cell receives an exact copy. Some minor changes should also occur at the rate of one error in a billion.

Although these speculations are quite sound, unfortunately most biologists concluded that protein is the genetic material. Proteins have necessary complexity and structure to form a long polymer. The twenty naturally occurring amino acids can be combined in almost unlimited variety, creating thousands and thousands of different types of proteins virtually. On the other hand, DNA was believed to be relatively small, invariant molecules with identical molecular weights. Each molecule of DNA was thought to be exactly the same as every other molecule of DNA, thus it did not have the variability required of the genetic material. The presence of DNA in chromosomes was attributed to structural functions.

Although the hypothesis that genes are made of protein was very widely held in the first half of the century, there was no solid experimental proof to support their view. But experiments carried for the identification of genetic material was suggesting DNA as genetic material rather than proteins. Experiments carried out by Griffith and Avery, Macloed, and McCarthy radically different in their design eventually led to the conclusion that genetic material is DNA and not protein.

Griffith's Experiment

In 1928, Frederick Griffith reported that heat-killed bacteria of one type could "transform" living bacteria of a different type. Griffith demonstrated this transformation using two strains of the bacterium *Streptococcus pneumoniae*. One strain produced smooth colonies on media in a petri plate. These strains produce a slimy coating consisting of polysaccharide.

This polysaccharide is responsible for protection of the strain during adverse conditions. It causes a fatal pneumonia. Another strain (R) produces rough colonies on media in a petri plate. This strain does not produce any slimy layer of polysaccharide. This strain does not cause any disease. Griffith conducted four experiments (Figure 2.1) involving smooth strain (S) and rough strain (R) of *S. pneumoniae* and mice. In set one experiment, he injected mice with live smooth strains (S) of *S. pneumoniae* and observed that mice died of pneumonia. In another set he injected live rough strain (R) of *S. pneumoniae* and observed that mice did not contract pneumonia. In the third set, he injected smooth strain (S) of *S. pneumoniae2* which are killed by heating at 60°C for 3 hours and observed that mice did not contract pneumonia.

This experiment indicates that the bacteria was dead and did not have any virulence further. Finally, a sample of heat-killed smooth strain (S), incapable of causing the disease, was mixed with live avirulent rough strain (R) of *S. pneumoniae* and the mixture was injected into a mice. This time a wholly unexpected result was obtained. The mice contracted pneumonia instead of being healthy. In addition, from this group, a large number of live smooth strain (S) of *S. pneumoniae* were isolated.

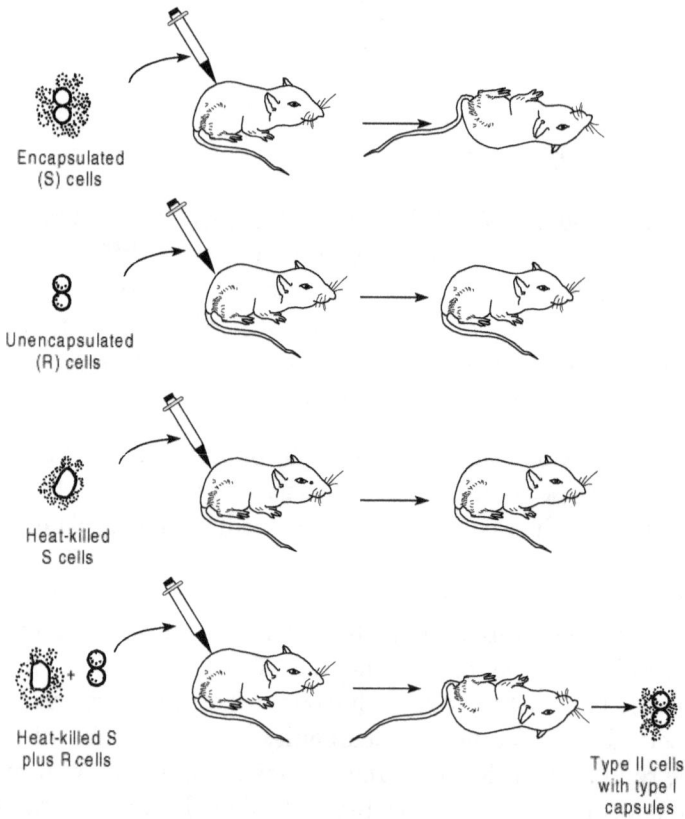

Encapsulated
(S) cells

Unencapsulated
(R) cells

Heat-killed
S cells

Heat-killed S
plus R cells

Type II cells
with type I
capsules

Figure 2.1   Diagrammatic representation of Griffith's experiment

The experiment result suggests that a component of the heat-killed smooth strain got (some transformation of pneumonia-causing agent) transferred into rough strain. Importantly, the live virulent bacteria obtained at the end of this experiment were S-type. The explanation cannot be that the type R in original inoculums regained their ability to synthesize their capsular polysaccharide. If this had happened then the resulting bacteria would have been type R, instead of smooth.

Avery and McCarthy Experiment

These experiments provided the first evidence that DNA was the genetic material. Avery and his associates prepared the filtrate from the heat-killed smooth strain of *S. pneumoniae* and then subjected it to specific degradation enzymes separately. The filtrate was divided into 3 parts, first part was exposed to protease enzyme which degrades protein, second part was exposed to ribonuclease enzyme which degrades the RNA and the last part was exposed to DNase enzyme which degrades DNA. After enzymatic treatment, the filtrate was tested for retention of its transforming ability. Surprisingly protease and RNases treatment had no effect on the transforming activity. However, digestion of the DNA with deoxyribonuclease totally destroyed the transforming ability, that the filtrate was not able to transform R type to S type. Gradually it became clear to Avery and her scolleagues that the genetic material is DNA and not protein. To confirm further, McCarthy isolated DNA from S-type cells of pneumonia and purified it by using electrophoresis and ultracentrifugation. Then by using direct physical and chemical analyses she showed the transformation substance to be DNA. Unfortunately, these findings did not establish the fact that DNA is genetic material. Most biologists believed that protein contamination in the preparation is responsible for transformation.

Hershey–Chase Experiment

Alfred Hershey and Martha Chase used radiolabelling method to detect the genetic material. Radiolabelling involves attachment of a radioactive atom or marker to the molecule in question. Proteins contain amino acids cysteine and methionine, which contain sulphur. Proteins can be radiolabelled by using S^{35}. As DNA contains no sulphur it cannot be labelled with S^{35}.

Conversely DNA can be labelled with P^{32}, which will not be incorporated into proteins because proteins do not contain phosphorus. Once labelled, samples of DNA and protein can be distinguished from each other because the two radioisotopes, P^{32} and S^{35} emit radiation of different characteristic energy.

Hershey and Chase prepared a radioactive sample of T2 bacteriophage one in which protein was labelled with S^{35} and other in which DNA was labelled with P^{32}, by infecting T2 phage on *E. coli* bacterial lawn grown with S^{35}- and P^{32}-labelled nutrients (Figure 2.2). The labelled phages produced by this culture were used to infect the next, non-radioactive culture of *E. coli*. However, this time the infection procedure was interrupted a few minutes after inoculation by agitating the cells in a waring blender. These few minutes were long enough for the phage genes to enter the bacteria, but not enough for new bacteriophages to be synthesized to kill the bacteria. The culture was then centrifuged so that the relatively heavy bacterial cells containing the phage genes collected at the bottom of the tube, leaving the empty phage particles in suspension. Hershey and Chase discovered that over 80% of the P^{32} was present in the bacterial pellet. The bacteria were than allowed to continue through the infection process and produce new phages. Almost half of the P^{32}, but less than 1% of the S^{32}, was present in these new phage particles. Clearly, this experiment has proved that the genetic material was DNA and not protein.

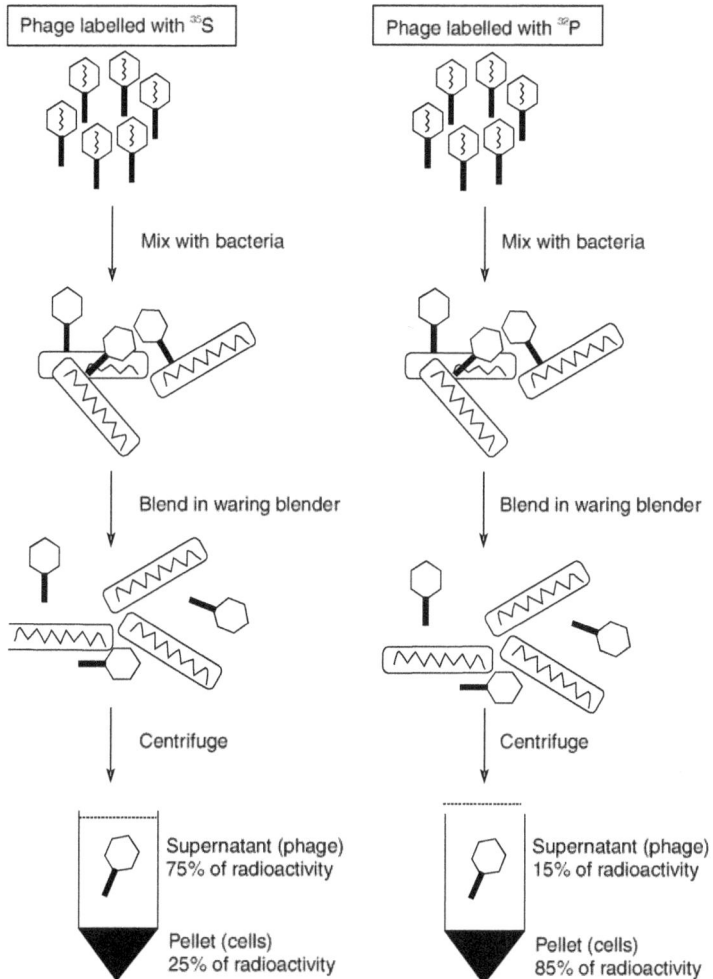

Figure 2.2 Diagrammatic representation of Hershey–Chase experiment

DNA STRUCTURE

At its most fundamental level, life is chemistry. All living things are made of molecules. Moreover, life is sustained by thousands of chemical reactions, and life would be impossible without these reactions. Therefore, in order to have a fundamental understanding of life and of the genetic material we must know something about its chemistry.

DNA molecule is a long polymer of nucleotides with nucleotides as basic units. Each nucleotide is composed of

1. a phosphate group
2. a five-carbon sugar or pentose
3. a cyclic nitrogen-containing compound called a base

Nitrogen Bases

These are rather complex single- or double- ring structures containing nitrogen and are attached to the 1′-carbon of the sugar. Nitrogen bases are of two types based on the number of rings (Figure 2.3).

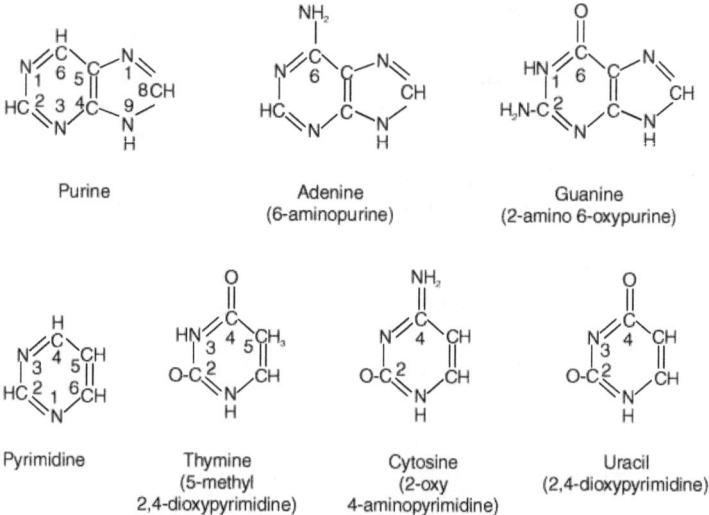

Purine	Adenine (6-aminopurine)	Guanine (2-amino 6-oxypurine)
Pyrimidine	Thymine (5-methyl 2,4-dioxypyrimidine)	Cytosine (2-oxy 4-aminopyrimidine)
		Uracil (2,4-dioxypyrimidine)

Figure 2.3 Molecular structure of nitrogen bases present in nucleic acid

Purines Purines are double-ringed nitrogenous bases which are linked to sugar by using nitrogen present at 9´-position. Adenine and guanine come under this group.

Pyrimidines Pyrimidines are single-ringed nitrogen bases which are linked to sugar by using nitrogen present at 1´-position. Thymine and cytosine come under this group.

Adenine	6-aminopurine
Guanine	2-amino 6-oxypurine
Cytosine	4-amino 2-oxypyrimidine
Thymine	2,4-oxy 5-methyl pyrimidine
Uracil	2,4-oxypyrimidine

Pentose Sugar

In DNA, the sugar component of the nucleotide is a pentose called 2´-deoxyribose base. Pentose sugars contain five carbon atoms and can exist in two forms (i) straight chain or Fischer structure and (ii) the ring or Haworth structure. In DNA, pentose sugar exists in ring form and is 2-deoxyribose type (Figure 2.4). The name 2-deoxyribose indicates that the standard ribose structure has been altered by replacement of the hydroxyl group (–OH) attached to carbon atom number 2´ with a hydrogen group (–H). The carbon atoms are always numbered in the same way with the carbon of the carbonyl group (–C=O), occurring at one end of the chain form numbered 1. It is important to remember the numbering of the carbons because it is used to indicate at which positions on the sugar other components of the nucleotide are attached.

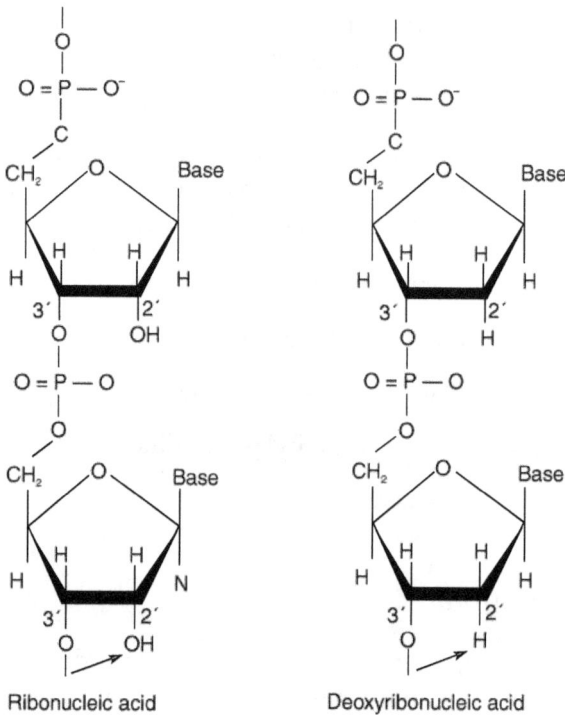

Figure 2.4   Pentose sugars present in nucleic acid

Phosphoric Acid Component

A molecule comprising the sugar joined to a base is called a nucleoside. This is converted into a nucleotide by attachment of a phosphoric acid group to the 5′-carbon of the sugar. Phosphate group is attached by phosphodiester bond. An ester is an organic compound formed from an alcohol and acid. In the case of a nucleotide, the alcohol group is the 5′-hydroxyl of the sugar and the acid is phosphoric acid. Hence we call the ester as phosphodiester, as a single phosphoric acid is linked to two sugars present on different nucleotides. Thus, the bond is called as phosphodiester bond. To be precise, we should call this bond as

3′–5′ phosphodiester bond so that there is no confusion about which carbon atoms in the sugar participate in the bond. Up to three individual phosphate groups can be attached in series giving a nucleotide monophosphate (NMP), nucleotide diphosphate (NDP) and nucleotide triphosphate (NTP). The individual phosphate groups are designated α, β and γ with the α-phosphate being the one attached directly to the sugar.

WATSON-CRICK MODEL OF DNA

Watson and Crick proposed the DNA model by using all the information that was available at that time. They used the data obtained from experiments carried out on DNA by Chargaff, and Maurice Wilkins and Rosalind Franklin. Before we go through Watson–Crick double helix we must look at the work of these people.

Chargaff rule   Chargaff rule states that the number of purines is always equal to the number of pyrimidines in a given DNA. The relationship is that the number of adenine residue equals the number of thymine, and the number of guanines equals the number of cytosine that is A = T and G = C. Thus we can say A + G = C + T.

X-ray diffraction studies   X-ray diffraction studies by M. Wilkins and R. Franklin suggested that DNA could be a helix with two regular periodicities of 3.4 Å and 34 Å along the axis of the molecules.

Based upon the above facts, Watson and Crick proposed the famous DNA structure model (Figure 2.5). The important features of this model are:

1. The DNA molecule is a double helix with single polynucleotides (phosphates, sugar, base) running in opposite directions.

2. The double helix is right-handed. This means that if the double helix is a spiral staircase that you were climbing up, the base would be on your right hand side.

3. The double helix has two different grooves. The helix is not absolutely regular. A major (~22Å) and a minor (~12Å) groove can be distinguished. This feature is important in the interaction between the double helix and the proteins involved in DNA replication and in expression.

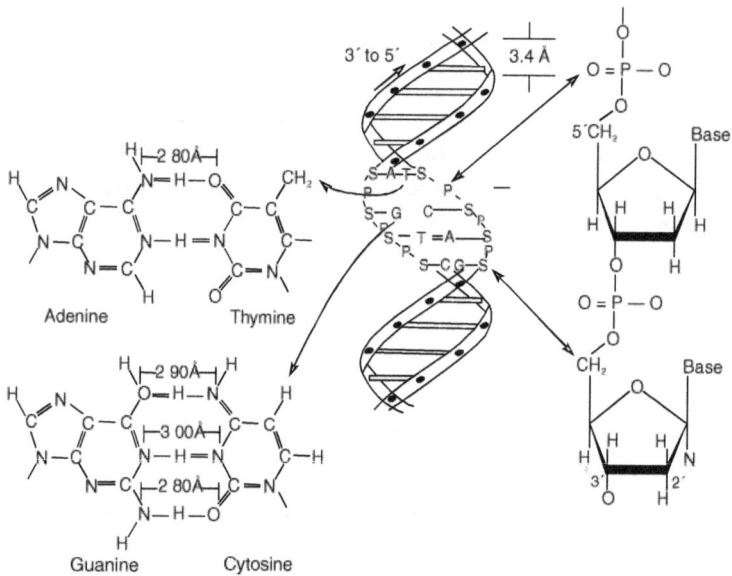

Figure 2.5 Molecular structure of DNA as proposed by Watson and Crick

4. The nitrogenous bases are stacked towards the inside of the helix. The experimental evidence also indicated that the sugar–phosphate backbone of the molecules is on the outside, with the bases inside the helix.

5. Bases of the two polynucleotides interact by hydrogen bonding. This is the explanation of Chargaff's base pairs.

An adenine residue in one of the polynucleotides is always adjacent to a thymine in the other strand; similarly guanine is always adjacent to cytosine. These two pairs of bases, and no other combinations are able to form hydrogen bonds between each other. These hydrogen bonds are the only attractive forces between the two polynucleotides of the double helix and serve to hold the structure together.

6. Ten base pairs occur per turn of the helix. The double helix executes a turn every ten base pairs (abbreviated as 10 bp). The height or pitch of the helix is 34 Å. The bases are stacked one on top of the other like a pile of plates. The space between the two base pairs is 8.4 Å and has an angle of 36°.

7. The diameter of the helix is 20 Å.

Conformational Flexibility of DNA Molecule

DNA is not a fixed and static invariant molecule. On the contrary, DNA molecules exhibit a considerable amount of conformational flexibility. These alternative forms of DNA have received increasing attention in recent years as it has been shown that the nucleotide sequence is one of the factors that influence the form taken by a segment of the double helix. The structure for DNA in a fibre is produced at a very high relative humidity (92%). This is called the B-form of DNA. The vast majority of the DNA molecules present in the aqueous protoplasm of living cells almost certainly exist in the B-form. The structure of DNA molecules change as a function of their environment. The exact conformation of a given DNA molecule or segment of DNA molecules will depend on the nature of the molecule with which it is interacting and amount of water content.

A-form   A-form of DNA is formed when relative humidity is 70%. It is right-handed, wider, incorporates more base pairs per helical turn and is less flexible than the B-DNA. The major groove is deep and narrow, whereas the minor groove is broad and shallow. The diameter of the DNA is 23 Å. Eleven base pairs are present per turn with a distance of 2.6 Å between them. A-form of DNA is less soluble than the B-form. The helix height or pith is 31Å.

Z-form Alexander Rich and his colleagues discovered in 1979 that DNA does not always have the right-handed form, it can be left-handed also. They showed that double-stranded DNA containing strands of alternating purines and pyrimidines (e.g. poly GC) can exist in an extended left-handed helical form. Because of the zigzag look of this DNA backbone when viewed from the side, it is often called as Z-DNA. The major groove is flat and the minor groove is narrow and very deep. The diameter of the DNA is 18 Å. Twelve base pairs are present per turn with a distance of 3.7 Å between them. Z-DNA structures tend to form in torsionally stressed DNA and are stabilized by dehydration. Originally, it was thought that Z-DNA would not prove of interest to biologists because it required very high salt concentration to become stable. However, it was found that Z-DNA can be stabilized in physiologically normal conditions, if methyl groups are added to the cytosine. Z-DNA may be involved in regulating given expression in eukaryotes.

Multiple-stranded DNA Under natural conditions, single-stranded RNA and double-stranded DNA are the rule. However, under laboratory conditions, it is possible to induce a third strand of DNA to interdigitate itself into the major groove of the double helix of normal DNA in a sequence-specific fashion. The binding is site-specific and rules of binding are less precise than normal.

Triple-stranded nucleotide chains were first created in 1957 by Alexander Rich, David Daves and Gary Felsenfeld, while they were creating artificial nucleic acids. At that time, triple-stranded DNA seemed like a laboratory curiosity. Now it seems of interest, because it may have valuable uses, both experimentally and clinically. Triplex DNA is generated by binding a known single-stranded DNA because single strand of DNA is capable of recognizing a relatively long sequence of the double-stranded DNA in a chromosome (Figure 2.6). Thus it is possible to selectively locate a particular gene loci or sequence. The second use of triplex DNA is to cut DNA at a specific place by adding a cleaving compound to both ends of the third strand of DNA. However, it seems to have good potential for therapeutic use and to help in studying and mapping the human genome.

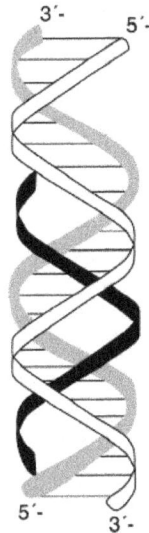

Figure 2.6   Triple helix DNA

More recently, four-stranded DNA molecules have been found in which double helices of certain sequences interdigitate to form four-stranded structures. Guanine can form base tetrads and DNA-containing runs of guanosine residues can form quadruplex structures, which may contribute to telomerase structure. Telomeric DNA consist of short, tandemly repeated sequence. These have been characterized from a number of eukaryotes and are generally GC-rich with guanine residues clustered on the one strand and cytosine residues on the other. They may form unusual quadruplex structures by unorthodox interactions between guanosine residues and which may play a role in protecting the telomere from end-joining reactions. Four-stranded DNA plays an important role in crossover.

RNA STRUCTURES

The nucleic acid, other than DNA, which exists in both prokaryotes and eukaryotes is ribonucleic acid. In some plant and animal viruses, RNA molecules can act as genetic material. Like DNA, RNA is also a long polymer of nucleotide. But there are some differences between DNA and RNA (Figure 2.7).

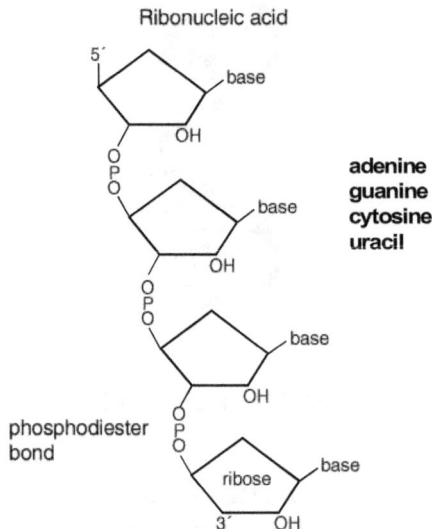

Figure 2.7   RNA structure

1. RNA is single-stranded and does not have antiparallel or complementary strand.

2. RNA also contains four major bases—adenine, guanine, cytosine and uracil instead of thymine.

3. The essential difference between DNA and RNA is the type of sugar each contains—RNA contains the sugar D-ribose (hence ribonucleic acid, RNA). Ribose sugar contains OH at 2′-carbon, which is absent in deoxyribose. This minor structural difference confers very different chemical and physical properties upon DNA and RNA. RNA is much stiffer due to steric hindrance and more susceptible to hydrolysis in alkaline conditions.

TYPES OF RNA

RNA molecules are divided into two groups based upon the function and abundance in the cell and this includes the major class and minor class.

Major Class of RNA

Based upon the function of RNA, the major class of RNA include the following types—mRNA, tRNA, rRNA and hnRNA.

Ribosomal and transfer RNAs are end products of gene expression and perform their role in the cell as RNA molecules. Messenger RNA, on the other hand, undergoes the second stages of gene expression and translation and has no function beyond acting as the intermediate between a gene and its final product, a polypeptide. Ribosomal and transfer RNAs are sometimes referred to as stable RNA, indicating that these molecules are long-lived in the cell. Both types of stable RNA (rRNA and tRNA) are involved in protein synthesis even though they are not themselves translated.

Ribosomal RNA   Ribosomal RNA molecules are components of ribosomes, the large multimolecular structures that act as factories for protein synthesis. During translation, ribosomes attach to mRNA molecules and migrate along them, synthesizing polypeptides as they go, analogous in a way to the role of RNA polymers in transcription. Ribosomes are made up of rRNA molecules and protein and are abundant in most cells (Figure 2.8).

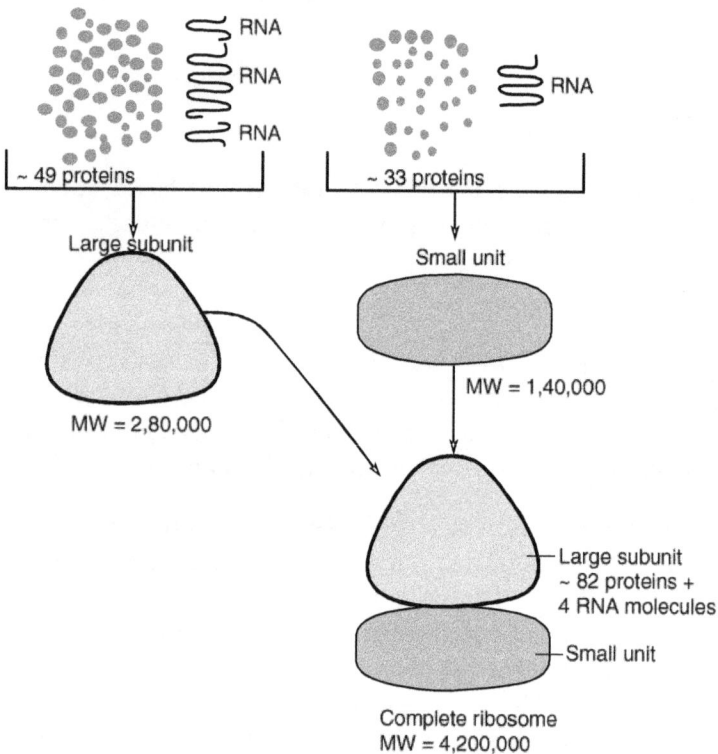

Figure 2.8   Components of ribosomes

Prokaryotic ribosome has a total molecular mass of 25,20,000 daltons. Prokaryotic ribosome is made up of two subunits with sedimentation coefficients of 50S and 30S. The large subunit contains two rRNA molecules of 23S and 5S together with 34 different polypeptides. The smaller subunit has a just a single 16S rRNA plus 81 polypeptides.

Eukaryotic ribosomes are also made of two subunits but in this case the sizes are 60S and 40S. The large subunit has three rRNAs (28S, 5.8S and 5S) and 49 polypeptides. The small subunit has a single 18S rRNA and 33 polypeptides. The additional rRNA of the eukaryotic large subunit is the 5.80 molecule, which in *E. coli* is present as an integral part of the 23S RNA.

Structure   The traditional view of ribosomal structure is that the rRNA molecules act as scaffolding to which the proteins, functional activity of the ribosome, are attached. To fulfil this role the rRNA molecules must be able to take up a stable 3D structure. This is achieved by inter- and intramolecular base pairing, with different rRNAs of subunit base pairing in an ordered fashion with each other and also more importantly with different parts of themselves. The traditional view of the rRNA as the scaffolding and the proteins as attachments that provide the real biological activity of the ribosomes is now being challenged by the latest exciting ideas about the function of rRNA. For many years it was believed that enzymatic catalysis is uniquely a feature of proteins and that RNA molecules cannot act as enzymes in *Tetrahymena* rRNA, self-splicing intron was discovered by Thomas each of the University of Colorado in 1980 (Figure 2.9). It was the first known example of a ribozyme and caused quite a stir: many biochemists unwilling to believe that RNA could have enzymatic activity as this was thought to be a property displayed only by proteins.

Synthesis of rRNAs   Single ribosome contains one copy of each of the different rRNA molecules, three rRNAs for the prokaryotic ribosome or five for the eukaryotic version. The most efficient system would be for the cell to synthesize equal numbers of each of these molecules. Synthesis of equal numbers of each rRNA molecules is assured by having an entire complement of rRNA molecules transcribed together as a single unit. Thus, the primary transcript is a long RNA precursor, the pre-RNA containing each rRNA separated by short spacers. The spacers are removed by processing events that release the mature rRNAs. A similar series of events brings about the synthesis, of eukaryotic rRNAs with the exception that only the 28S,18S and 5.8S genes are transcribed together. The 5S RNA genes occur elsewhere on the eukaryotic chromosomes and are transcribed independent of the main transcript. A part from the only synthesis and removal of rRNA from Pre-rRNA occurs, a second type of processing called. Chemical modification also occurs. Certain nucleotides in the rRNA undergo conversion to unusual forms of nucleotides by alteration in their chemical structures. In eukaryotic rRNAs the commonest form of modification is 2′–O-methylation, in which the hydrogen of the –OH group attached to the 2′-carbon is replaced by a methyl group. These modifications do not occur at random, rather they occur at specific position, i.e., each copy of 28S rRNA is modified at exactly the same nucleotide positions.

As single rRNA gene cannot meet the demand of rRNA synthesis, multiple copies of rRNA genes exist. In *E. coli,* 7 copies of genes exist. In eukaryotes there can be an even greater demand for rRNA synthesis, hence 50–5000 identical copies of the rRNA transcription units are present depending upon the species. In eukaryotes, these rRNA transcription units are arranged one after the other, separated by non-transcribed spacer. In some instances, eukaryotes satisfy the demand for rRNA synthesis by using gene amplification strategy. This involves replication of rRNA genes into multiple DNA copies, which subsequently exist as independent molecules not attached to the chromosomes.

Transfer RNA Transfer RNA molecules (Figure 2.9) are also involved in protein synthesis but the part they play is completely different from that of rRNA. Transfer RNAs are in fact the adapter molecules that read the nucleotide sequence of the mRNA transcript and convert it into a sequence of amino acids. tRNA was first isolated (1959) and sequenced (1965) by Robert Holley. Transfer RNA molecules are relatively small, mostly between 74 and 95 nucleotides for different molecules in different species. Each organism synthesizes a number of different tRNAs, each in multiple copies. However all tRNAs take cloverleaf model (Figure 2.10) after synthesis, except those tRNA molecules present in mitochondria which have T-shaped structure. tRNA molecule which has got cloverleaf structure has 5 components.

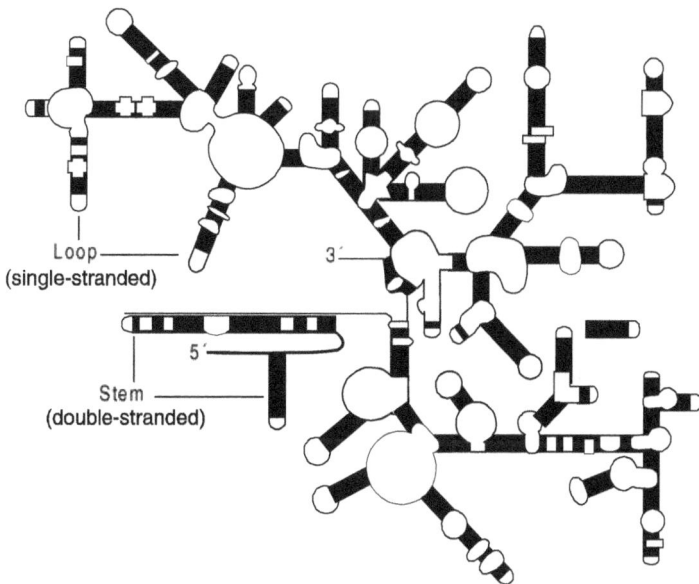

Figure 2.9 tRNA structure

Acceptor arm It is formed by a series of usually seven base pairs between nucleotides at the 5′ and 3′ ends of the molecule. During protein synthesis, an amino acid is attached to the acceptor arm of the tRNA. The amino acid becomes attached to the end of the acceptor arm of the tRNA cloverleaf. Each tRNA molecule forms a covalent linkage with its specific amino acid by a process called aminoacylation or charging which is catalysed by a group of enzymes called the aminoacyl-tRNA synthetases. In most cells there is a single aminoacyl-tRNA synthetase for each amino acid. For example, one enzyme can charge each member of a series of iso-accepting tRNAs. All tRNAs at the acceptor end have the sequence 5′–CCA–3′ at their end.

D or DHU arm　This arm invariably contains the modified pyrimidine called dihydrouracil.

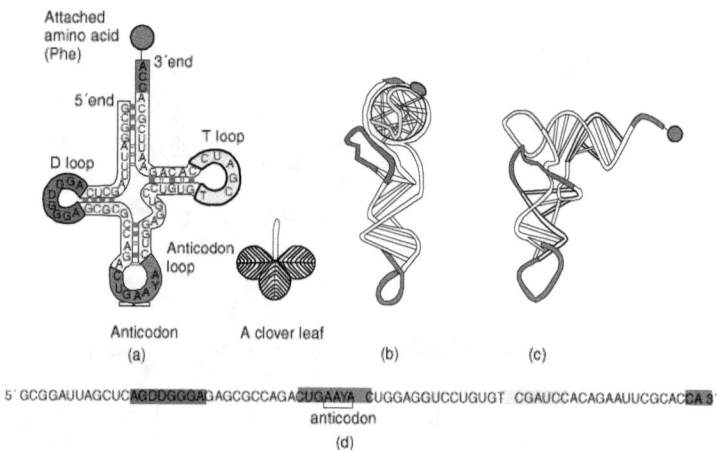

Figure 2.10　(a), (b), (c) Different models of tRNA structure representation (d) RNA sequence of tRNA molecule

Anticodon arm　It plays the central role in decoding the biological information carried by the mRNA. Codon recognition is a function of the anticodon loop of the tRNA, specifically of the trinucleotide called the anticodon. This trinucleotide is complementary to the codon and can therefore attach to it by base pairing. The specificity of the genetic code is therefore ensured because the anticodon present on a particular tRNA is one that is complementary to a codon for the amino acids with which the tRNA is charged.

Extra, optional or variable arm   It is a loop of just 3–5 nucleotides or 13–21 nucleotides.

T and C arm   It is the arm which contains a pseudouracil (another modified pyrimidine base) between T and C nucleotide.

Synthesis   In both prokaryotes and eukaryotes tRNA are transcribed initially as precursor-tRNA, which is subsequently processed to release the mature molecule. In *E. coli* and humans there are several separate tRNA transcription units, some containing just one tRNA gene and some with as many as seven different tRNA genes in cluster. A pre-tRNA molecule is processed by a combination of different ribonucleases that make specific cleavages at the 5 ′ (RNase-P) and 3 ′ (RNase-D) ends of the mature tRNA sequence.

Interestingly, in eukaryotes 5 ′–CCA–3 ′ sequence is not present in the tRNA gene at the expected position. This sequence is added after transcription by another processing enzyme called tRNA nucleotidyl transferase. In contrast, in prokaryotes the final CCA is more frequently coded by the tRNA gene and is therefore transcribed in the normal manner. This sequence (5 ′–CCA–3 ′) is removed by RNase-D during processing of the pre-tRNA and has to be replaced by a prokaryotic nucleotidyl transferase enzyme.

Like rRNA, tRNA also undergoes modifications. Over 50 types of chemical modifications have been discovered so far with tRNA nucleotides, each catalysed by different tRNA-modifying enzymes. The most common types are methylation base rearrangements, double-bond saturation, deamination, sulphur substitution or addition of more complex groups. The reasons for most of these modifications are unknown.

Minor Class of RNA

This group of RNA are produced in small amounts and they are not translated nor excessively modified.

snRNA or U-RNA   These RNA molecules are low-molecular weight compound and found in the nucleoplasm. These RNA molecules are rich in uridine residues. Hence they are also called U-RNA. These RNA molecules play an important role in RNA splicing in combination with other proteins.

Till now five snRNA have been discovered—U1, U2, U4, U5 and U6. These are located in the nucleus. Another U3 has also been discovered which is present in nucleolus. snRNA's size ranges from 100 bases to 215 bases. These snRNAs do not exist as free RNA molecules, but as small nuclear RNA–protein complex called snRNPs or small nuclear ribonucleic proteins. However, the exact protein composition of intact spliceosomes is not established. About 10^6 copies of snRNA genes are present per cell. snRNA are very stable and are capped by trimethyl guanosine cap.

lRNA   These are short sequences used as primers for lagging strand DNA synthesis.

scRNA   These are called as small cytoplasmic RNA. They are low molecular weight RNA molecules found in cytoplasm with various functions, e.g. pRNA (Prosomal RNA) a small RNA associated with approximately 20 proteins and found packaged with mRNA in the mRNP.

Telomerase RNA   A nuclear RNA which contains the template for telomere repeats and forms part of the enzyme telomerase.

Antisense RNA   Antisense RNA is complementary to mRNA and can form a duplex with it to block protein synthesis. Naturally occurring antisense RNA is found in many systems but predominantly in bacteria. These are also called as mRNA-interfering complementary RNA or mic RNA.

REVIEW QUESTIONS

1. Describe in detail Griffith experiment, Avery and McCarthy experiment, and Hershey and Chase experiment.

2. Describe the Watson and Crick model of DNA.

3. Describe in detail the chemical molecules found in DNA.

4. Explain the structure of A and Z types of DNA.

5. Describe the structure of RNA.

6. Write in detail about ribosomal RNA.

7. Describe the structure of tRNA in detail.

8. Describe different types of RNA that are found in minor class.

3

DNA REPLICATION
AND REPAIR

The ability to reproduce is one of the most fundamental properties of all living systems. This process of duplication can be observed at several levels; organisms duplicate by cellular division; the gametic material duplicates by cellular division; and the genetic material duplicates by replication. The machinery that cells use for replication is also called as DNA replication. The capacity for self-duplication is presumed to have been one of the first critical properties to appear on the path towards evolution of the first primitive life forms. The first carriers of genetic information were probably RNA molecules that possessed the capability for self-replication. As evolution progressed and RNA molecules gave way to DNA molecules, the process of replication became considerably more complete, requiring a large number of auxiliary material. Thus although a DNA molecule contains the information for its own duplication, it lacks the ability to perform the activity itself. Nevertheless, molecular biologists in the early 1950s were uncertain about the overall pattern of the process. Three different strategies for replication of the double helix seemed possible.

1. *Semiconservative replication*   In this method, each daughter molecule contains one polynucleotide derived from the original molecule and one newly synthesized strand.

2. *Conservative replication*   In this method, one daughter molecule contains both parent polynucleotides and the other daughter contains both newly synthesized strands.

3. *Dispersive replication*   In this method, each strand of each daughter molecule is composed partly of the original polynucleotides and partly of newly synthesized polynucleotides.

MESELSON–STAHL EXPERIMENT

In 1958, M. Meselson and E. Stahl reported the results of an experiment designed to determine the mode of DNA replication.

The researchers determined the density of the strands using a technique known as density-gradient centrifugation. In this technique, a caesium chloride (CsCl) solution is spun in an ultracentrifuge at high speed for several hours. If DNA (or any other substance) is added, it concentrates and forms a band in the tube at the point where its density is the same as that of the CsCl. If several types of DNA with different densities are added, they form several bands. The bands are detectable under UV light at a wavelength of 260 nm, which nucleic acids absorb strongly.

Meselson and Stahl grew *E. coli* in a medium containing a heavy isotope of nitrogen (N^{15}). After growing for several generations on the N^{15} medium, the DNA of *E. coli* was denser. Then the culture was spun in a low-speed centrifuge, the heavy medium was discarded and the bacteria resuspended in a medium containing only N^{14}. The bacteria were then allowed to undergo one round of cell division, which takes roughly 20 min. for *E. coli* during which time each DNA molecule replicates just once. Some cells were then taken from the culture, their DNA was purified and a sample was analysed by density gradient. The result showed a single band of DNA at a position corresponding to a buoyant density intermediate between the values expected for N^{15} and N^{14} (Figure 3.1). DNA double helix contains roughly equal volumes or amounts of N^{15} polynucleotides. These results at this moment eliminate the conservative replication model.

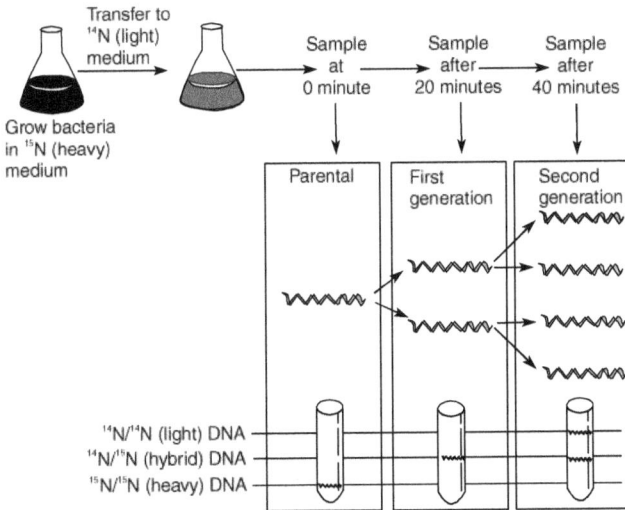

Figure 3.1 Diagrammatic representation of Meselson and Stahl experiment

To distinguish between semiconservative and dispersive replication, *E. coli* culture was incubated for 40 min. instead of 20 minutes, so that two rounds of replication occurs. Again the cells were removed and their DNA was analysed in a density gradient. Two bands appeared, one representing the same hybrid molecules as before, the second corresponding to wholly N^{14} DNA. These results suggest that DNA replication is by semiconservative mode. In contrast, the dispersive mode can be discarded because that method would still produce only hybrid molecules and in fact would continue to do so for a large number of cell generations (Figure 3.2).

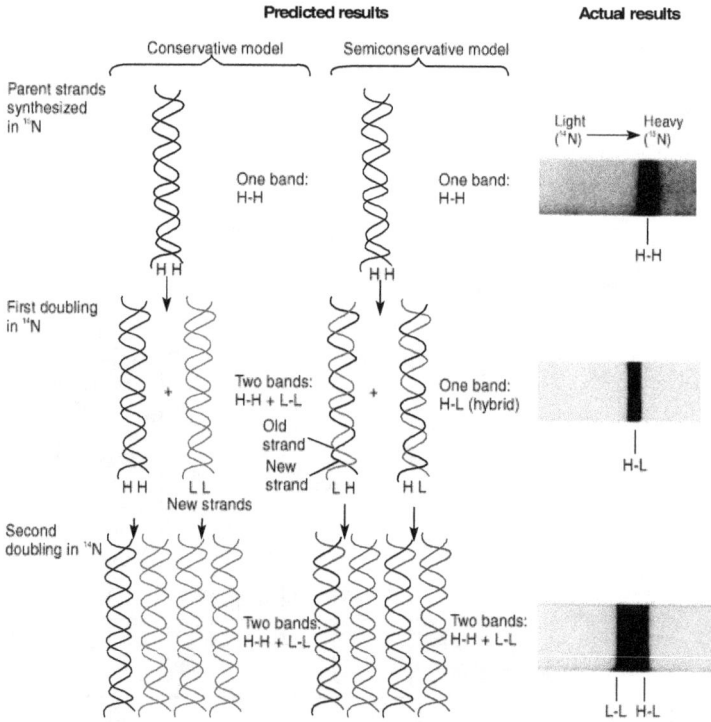

Figure 3.2 Theoretical explanation for Meselson and Stahl experiment

ENZYMOLOGY OF DNA REPLICATION

Like all metabolic processes, DNA replication is also under the control of enzymes. Physical, chemical, biochemical and mutation studies indicate that DNA replication is under the strict control of enzymes.

DNA Polymerase

DNA polymerase enzyme is responsible for constructing new DNA strands. DNA polymerase enzyme catalyses the addition or synthesis of new complementary strands using a template strand of DNA and all four deoxyribonucleotide triphosphates. There are three major DNA polymerase enzymes that will polymerize nucleotides into a growing strand of DNA in *E. coli*. These enzymes are DNA polymerase I, II and III. DNA pol I and II are primarily used for DNA repair and for replicating a small length or segment of DNA. All the DNA polymerase that have been studied have two basic requirements:

1. A template DNA strand to copy.

2. A primer strand to which nucleotides can be added.

Likewise all DNA polymerase enzymes have two basic properties:

1. 5′–3′ polymerization capacity

2. 5′–3′ and exonuclease activity

The 5′–3′ exonuclease activity of DNA pol I differs from the 3′–5′ exonuclease used by both DNA pol I and pol III in two different/important ways. Firstly 5′–3′ exonuclease can remove one nucleotide at a time from a region of DNA that is improperly base-paired. The nucleotides it removes can be either ribonucleotides or deoxyribonucleotides. Secondly, 5′–3′ exonuclease can also remove groups of altered nucleotides in 5′–3′ direction, removing from one to ten nucleotides at a time. This ability is important in the repair of damaged DNA. A typical bacterial cell contains approximately 300 to 400 molecules of DNA pol I per cell and only about 40 copies of DNA polymerase II and 10 copies of DNA pol III. Eukaryotes have 9 types of DNA polymerases named α, β, γ, δ, ε, ξ, η, θ and τ. DNA polymerase δ seems to be the major replicating enzyme in eukaryotes.

Helicases

Helicase is the enzyme responsible for unwinding of the double-stranded DNA into single-stranded DNA by breaking the hydrogen bonds. Helicase works along with topoisomerase. Helicase uses ATP as the energy source for unwinding. It contains an ATPase that breaks down two molecules of ATP for every base pair separated.

SSB

Single-stranded DNA-binding proteins or helix destabilizing proteins, bind only to single-stranded DNA. The SSB proteins are not enzymes, but rather serve to shift to equilibrium between double- and single-stranded DNA in the direction of the single-stranded species. These proteins not only keep the two strands of DNA separated in the area of the replication origin, but also protect the DNA from nucleases that cleave single-stranded DNA. SSB proteins bind to DNA as tetramers. First a single SSB binds, then it stimulates other SSB to bind and form tetramer. SSB proteins not only maintain the DNA in single-stranded nature but are also involved in removing the secondary structures, inhibiting the endonuclease activity and enhancing the replication rates.

Topoisomerase

This enzyme is responsible for the prevention of torsion force created on DNA due to helicase action. Topoisomerase can release a supercoiled DNA but cannot coil a DNA. Topoisomerase may be divided into two classes depending upon the cleavage mode.

Type I DNA topoisomerase This enzyme is responsible for prevention of torsion force created on DNA. It reversibly cuts a single strand of the double helix. It has both nuclease and ligase (strand-resealing) activities. This enzyme does not require ATP, but rather appears to store the energy from the phosphodiester bond it cleaves reusing the energy to reseal the strand. By creating a transient "nick", the DNA helix on either side of the nick is allowed to rotate at the phosphodiester bond opposite the nick, thus relieving accumulated supercoils. Type 1 topoisomerase relaxes negative supercoils (that is, those that contain fewer turns of the helix than relaxed DNA) in prokaryotes. But it can cleave both negative and positive supercoils (that is, those that contain more turns of the helix than relaxed DNA) in eukaryotic cells.

Type II DNA topoisomerase This enzyme binds tightly to the DNA double helix and makes transient breaks in both strands. The enzyme then causes a second strand of the DNA double helix to pass through the break and finally reseals the break. As a result, both negative and positive supercoils can be relieved. Type II DNA topoisomerase is required in both prokaryotes and eukaryotes for the separation of interlocked molecules of DNA following chromosomal replication. This topoisomerase also does not require ATP.

DNA Gyrase

A type II topoisomerase found in *E. coli* has the unusual property of being able to introduce negative supercoils into resting circular DNA. This facilitates the future replication of DNA because the negative coils neutralize the positive supercoils introduced during opening of the double helix.

DNA Ligase

This is the enzyme responsible for the ligation of nascent chain and Okazaki fragments. It joins the DNA strand whenever there is a break. The joining of these two stretches of DNA requires energy, which in humans is provided by the cleavage of ATP to AMP + PPi.

Primase

A specific RNA polymerase, called primase, synthesizes short stretches of RNA that are complementary and antiparallel to the DNA template. In the resulting hybrid duplex, A pairs with U and G pairs with C. These short RNA sequences are constantly being synthesized at the replication fork on the lagging strand but very few are required on the leading strand.

Primosome

Prior to the beginning of RNA primer synthesis on the lagging strand, a pre-priming complex consisting of 6 proteins is formed. The complex along with the primase enzyme is called as primosome.

DNA A Proteins

Twenty to fifty monomers of DNA A proteins bind to specific nucleotide sequences at the origin of replication, which is particularly rich in A.T base pairs. This ATP-requiring process causes the double-stranded DNA to melt, i.e., to convert from double-stranded to single-stranded nature.

STEPS IN PROKARYOTIC DNA REPLICATION

When the two strands of DNA double helix are separated, each can serve as a template for the replication of a new complementary strand, producing two daughter molecules each of which contains two DNA strands with an antiparallel orientation. The enzymes involved in DNA replication process are template-directed polymerases that can synthesize the complementary sequence of each strand with extraordinary fidelity.

In order for the two strands of the parental double helical DNA to be replicated, they must first separate or ("melt"), at least in a small region because the polymerase uses only single-stranded DNA as a template. In prokaryotes, DNA replication begins at a single, unique nucleotide sequence called the origin of replication or *ori* site. At the origin of replication, first DNA B/A protein binds, then *oriC* protein binds leading to the formation of a complex spanning about 150–250 bp. This complex is called as origin replication complex. This complex leads to the local denaturation and unwinding of an adjacent A + T-rich region of DNA. The interaction of proteins with *ori* defines the start site of replication and provides a short region of ssDNA essential for initiation of synthesis of the nascent DNA strand. Then helicase binds and allows processive unwinding of double-stranded DNA into single-stranded DNA. As helicase unwinds the DNA, DNA-binding (SSB) proteins bind and stabilize the single-stranded DNA.

As the two strands unwind and separate the DNA, they form a 'V' shaped structure or region called replication fork (Figure 3.3). A replication fork consists of four components that form in the following sequence:

1. DNA helicase unwinds a short segment of the parental duplex DNA.
2. A primase initiates synthesis of an RNA molecule that initiates DNA synthesis by acting as primer.

3. The DNA polymerase initiates daughter strand synthesis.

4. SSBs bind to ssDNA and prevent premature re-annealing. It moves along the DNA as synthesis occurs. Some people define replication fork as the site of active DNA synthesis.

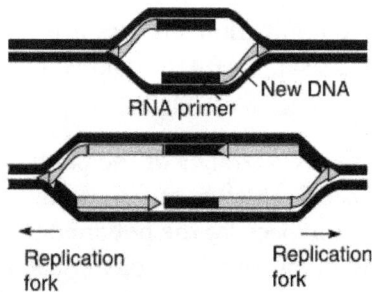

Figure 3.3 Replication fork

The polymerase III holoenzyme binds to template DNA as a part of a multiprotein complex that consists of several polymerase accessory factors. DNA polymerase synthesizes DNA only in the 5′ to 3′ direction and only one of the several different types of polymerases is involved at the replication fork. As the DNA strands are antiparallel, the DNA polymerase functions asymmetrically. On the leading (forward) strand, the DNA is synthesized continuously. On the lagging strand (retro strand) the DNA is synthesized in short (1–5 kb) fragments. These DNA fragments are called as okazaki fragments.

DNA polymerase present in the replication fork shares three important properties (i) chain elongation (ii) Processivity and (iii) Proofreading. Chain elongation accounts for the rate at which polymerization occurs. Processivity is expression of the number of nucleotides added to the nascent chain before the polymerase disengages from the template. The proof function identifies copying errors and corrects them. Pol III is an enzyme with high processivity and catalysing capacity than others. The initiation of DNA synthesis requires priming by a short length of RNA about 10–200 nucleotides long. This priming process involves the nucleophilic attack by the 3'-OH group of the RNA primer on the α-phosphate of the deoxyribonucleoside triphosphate that enters first, with the splitting off of pyrophosphate. The 3'-OH group of the recently attacked deoxyribonucleoside monophosphate is then free to carry out a nucleophilic attack on the deoxyribonucleoside triphosphate that enters next. The selection of the proper deoxyribonucleoside whose terminal 3'-OH is to be attacked is dependent upon proper base pairing with the other strand of the DNA molecule, according to the rules proposed originally by Watson and Crick. By this stepwise process, the template dictates which deoxyribonucleoside triphosphate is complementary, and by hydrogen bonding holds it in place while the 3'-OH group of the growing strand attacks and incorporates the new nucleotide into the polymer. The entire replication complex is depicted in Figure 3.4.

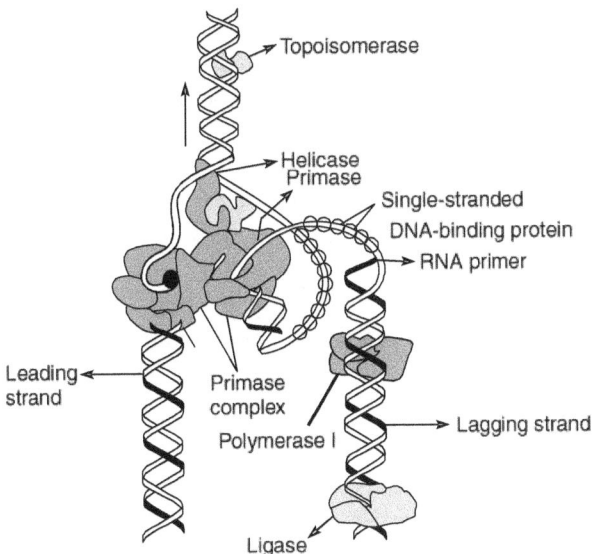

Figure 3.4   Replication complex

DNA pol III not only incorporates a nucleoside. It makes sure that the added nucleoside is correctly matched to its complementary base on the template and edits its mistakes. If by mistake it incorporates a wrong nucleotide, then DNA pol III hydrolytically removes the misplaced nucleotide and replaces it with the correct nucleotide. To ensure that replication occurs without any problem, the helicase acts on the lagging strand to unwind dsDNA in a 5′ to 3′ direction. The helicase associates with the primase to afford the latter, proper access to the template. This allows the RNA primer to be made and the polymerase to begin replicating the DNA. This is an important reaction sequence since DNA polymerase cannot initiate DNA synthesis *de novo.* The mobile complex between helicase and primase is called as primosome. This process proceeds as long as the DNA polymerase does not come in contact with RNA. Once the polymerase encounters the RNA, the lagging strand DNA is released from the polymerase. But the polymerase remains attached to the replication fork complex proteins. The exact mechanism of this process is not clear yet. But research suggests, that the B subunit of DNA pol II carries out the function of releasing and re-clamping single-stranded DNA in an energy-dependent fashion.

After the entire replication completes, the RNA primers are removed by DNA pol II and fresh base pairs of deoxyribose sugars are added in its position. Then ligase joins the DNA molecules and the process of DNA replication proceeds towards termination. The termination of the replication of a circular chromosome, *E. coli* chromosome is circular, presents no major topological problems. At the end of the theta structure both the ′V′ structure junctions have preceded around the molecule. The region of termination on the *E. coli* chromosome, the terminus region, is 180° from *ori* site. There are six termination sites, the protein product of the *tus* gene arrests the replication of both the strands of DNA. One interesting aspect of the termination of *E. coli* DNA replication is that the cells are viable even if the whole terminator gene is deleted. One may question as to why RNA is used to prime DNA synthesis, where DNA can be used directly to avoid the exonuclease and resynthesis activity. This strategy is followed to have low error rate. Priming is an inherently error-prone process since nucleotides are initially added without a stable primer configuration. To prevent long-term errors in the DNA, an RNA primer is put in and it can later be recognized and removed thus making very few errors.

The other question that comes to the mind is why DNA synthesis cannot take place in the 3′–5′ direction. The answer has to do with proofreading and the exonuclease removal of mismatched nucleotides. When an incorrect nucleotide is found and removed, the next nucleotide brought in, in the 5′–3′ direction has a triphosphate available to provide the energy for its own incorporation. Let us consider a polymerase with the ability of adding nucleotides in the opposite direction. The energy for the phosphodiester bond would be coming from the triphosphate already attached in the growing 3′–5′ strand. Then if an error in complementarity were detected and if the polymerase removed the most recently added nucleotides from the 3′–5′ strand, the last nucleotide in the double helix would no longer have a triphosphate available to provide energy for the diester bond with the next nucleotide. Continued polymerization would thus require additional enzymatic steps to provide the energy needed for the process to continue. This could stop or slow down the process considerably. As it is, the process incorporates about four hundred nucleotides per second with an error rate of about one incorrect pairing per 10^9 bases.

EUKARYOTIC DNA REPLICATION

In eukaryotes, the process of DNA replication is the same as that of the bacterial/prokaryotic DNA replication with some minor modifications. In eukaryotes, the DNA molecules are larger than in prokaryotes and are not circular; there are also usually multiple sites for the initiation of replication. Thus, each eukaryotic chromosome is composed of many replicating units or replicons—stretches of DNA with a single origin of replication (Figure 3.5 and 3.6). In comparison, the *E. coli* chromosome forms only a single replication fork. In eukaryotes, these replicating forks, which are numerous all along the DNA, form "bubbles" in the DNA during replication. The replication fork forms at a specific point called autonomously replicating sequences (ARS). The ARS contains a somewhat degenerate 11-bp sequences called the origin replication element (ORE). The ORE is located adjacent to an 80-bp AT-rich sequence that is easy to unwind.

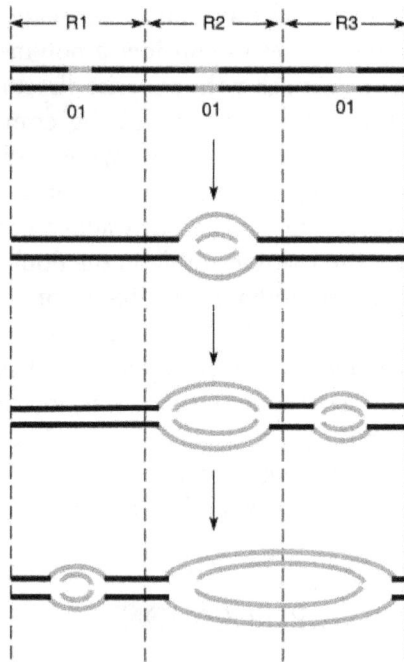

Figure 3.5   Multiple replication forks in eukaryotes

In yeast, ORE is called as DUE (DNA-unwinding element). Multiple origins allow eukaryotes to replicate their larger quantities of DNA in a relatively short time, even though eukaryotic DNA replication is considerably slowed by the presence of histone proteins associated with the DNA to form chromatin.

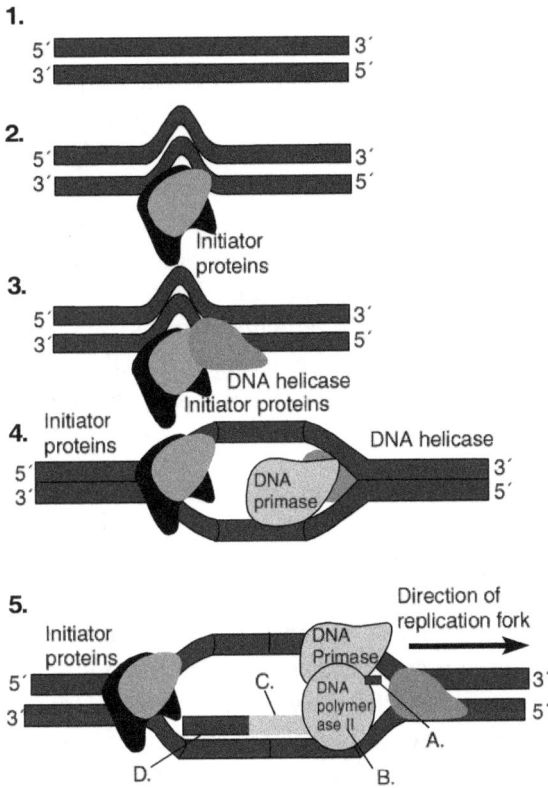

Figure 3.6 Eukaryotic replication fork

Eukaryotes have clamp-loader complex, similar to β subunit of pol of prokaryotes, and a six-unit clamp called the proliferating cell nuclear antigen. The RNA primers are removed during okazaki fragment completion by mechanisms similar to those in prokaryotes. In eukaryotes, RNase enzymes remove the RNA primers in okazaki fragments; a repair polymerase fills gaps and a DNA ligase forms the final seal.

Helicases, topoisomerase and single-strand binding proteins play roles similar to that in prokaryotes. All the enzymatic processes are generally the same in prokaryotes and eukaryotes. DNA replication developed in prokaryotes, and was refined as prokaryotes evolved into eukaryotes. The completion of the replication of linear eukaryotic chromosome involves the formation of specialized structures at the tips of chromosomes. Termination of replication in eukaryotes is different from that of prokaryotes. Eukaryotes have linear chromosome, and once the first primer is removed from the strand, there is no known way to fill in the gap, since DNA cannot be extended in the 3′–5′ direction and there is no 3′-end upstream as there would be in a circle. If this were actually the situation, the DNA strands would get shorter every time they replicated and genes would be lost forever.

To avoid this condition, the cell has devised a system. The ends of chromosomes do not have genes and instead, they are composed of many repeats of short, GC-rich sequences. The exact sequence of the repeat in a telomere is species-specific. These repeats are added to the 3′-end of DNA, not by semiconservative replication, but by an enzyme called telomerase. This enzyme has small RNA of 159–200 bp length which act as template. The telomerase adds many repeated copies of its characteristic sequence to the 3′ ends of chromosome. Priming for synthesis of the opposite strand can then occur within these telomeres. Interestingly, somatic cells lack telomerase while the germ cell retains the enzyme. Clearly, a picture of the "replication apparatus" of eukaryotic organisms is beginning to emerge, but still there are many things which need to be explored.

Rolling Circle Replication

This type of replication (Figure 3.7) is seen in some animal viruses such as lambda (λ) viruses. The replicating fork resembles that in *E. coli* DNA replication, with continuous synthesis on the leading strand and discontinuous synthesis on the lagging strands. One can actually imagine the replicating DNA as a roll of toilet paper unrolling as it speeds across the floor. In the rolling-circle mode of replication, a nick is made in one of the strands of the circular DNA, resulting in the replication of a circle and a tail. The unrolled part represents the growing double-stranded progeny DNA. The progeny DNA is several genomes long before it is packaged. These multiple-length DNAs are called as concatemers. This mode of replication is sometimes also called sigma mode replication because the intermediates resemble the lower case Greek letter sigma (σ).

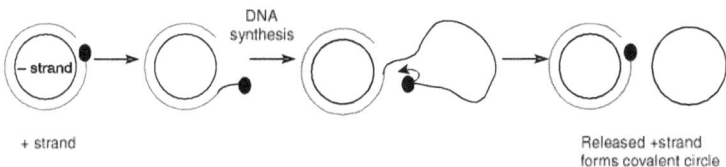

Figure 3.7   Rolling circle replication

DNA REPAIR

DNA is constantly being subjected to environmental interaction that causes the alteration or removal of nucleotide bases. Radiations, chemical mutagens, heat, enzymatic errors, and spontaneous decay constantly damage DNA. If the damage is not repaired, a permanent mutation may be introduced that can result in a number of deleterious effects including loss of control over the proliferation of the mutated cell. In the long evolutionary challenge to minimize mutation, cells have evolved numerous mechanisms to repair damaged or incorrectly replicated DNA. Many enzymes, acting alone or in concert with other enzymes, repair DNA. Most of the repair enzymes are involved in recognizing the lesion, excising the damaged section of the DNA strand and using the sister strand as a template, filling the gap left by the excision of the abnormal DNA (Figure 3.8).

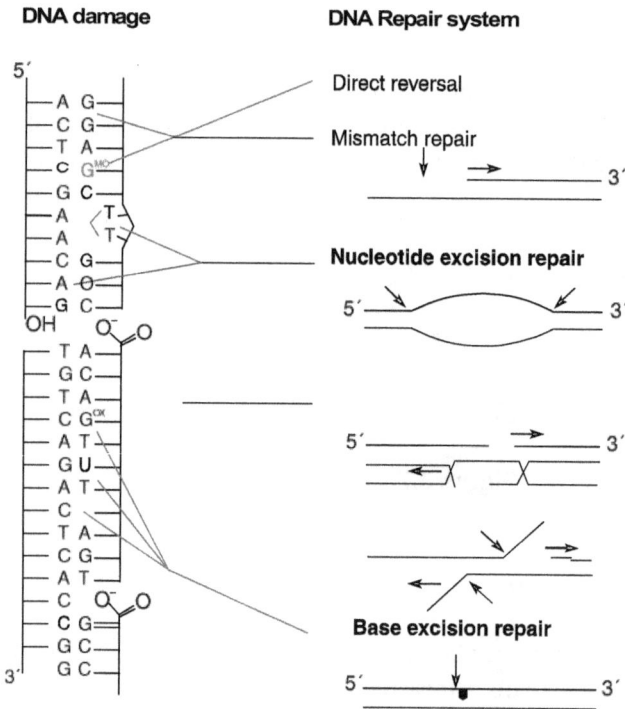

Figure 3.8   Different DNA repair systems

Repair systems are generally placed in four broad categories: damage reversal, excision repair, double-strand break repair and post-replicative repair. Enzymes of the repair system have been conserved during evolution. That is, enzymes found in *E. coli* have homologues in yeast, fruitflies, etc.

Mechanism of DNA Repair

Mechanisms for the repair of damaged DNA are probably universal. *E. coli* possesses at least three different mechanisms for the repair of DNA.

Direct repair Direct repair involves neither removal nor replacement of bases or nucleotides; rather the covalent modification in DNA is simply reversed. It may be of the following types.

Damage reversal or photo-reactivation Exposure of a cell to ultraviolet light can result in the covalent joining of two adjacent pyrimidines producing a dimer. Although cytosine–cytosine, cytosine–thymine dimers are also formed, the principal products of UV irradiation are thymine–thymine dimers (Figure 3.9). These thymine dimers prevent DNA polymerase from replicating the DNA strand beyond the site of dimer formation. In *E. coli*, an enzyme called DNA photolyase (deoxyribodipyrimidine photolyase or photo-reactivating enzyme) detects and binds to the damaged DNA site. Then the enzyme absorbs energy from visible light, which activates it so it can break the bonds holding the pyrimidine dimer together. The enzyme then falls free of the DNA. This enzyme thus reverses the UV-induced dimerization. Photo-reactivation or photo-restoration is a light-dependent DNA repair mechanism in which certain types of pyrimidine dimers are cleaved. This repair pathway is found in many prokaryotes and lower eukaryotes but absent in higher eukaryotes. Photo-reactivation should not be confused with other, non-enzymatic mechanisms of monomerization.

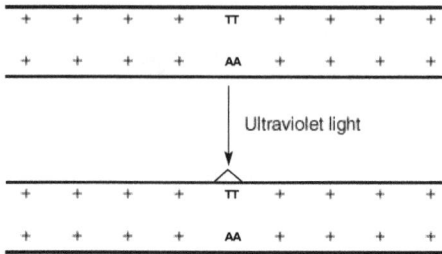

Figure 3.9   Thymine dimer structure

Direct photo-reversal   This occurs during continued UV–irradiation of DNA and reflects the establishment of equilibrium between the monomer and dimer states of adjacent pyrimidines. Sensitized photo-reversal occurs in the presence of tryptophan, which donates an electron to facilitate monomerization.

Excision repair   The percentage of mutations that can be handled by direct reversal is necessarily small. Most mutations involving pyrimidine dimers are handled by a different mechanism. Most of them are removed by a process called excision repair (Figure 3.10). Excision repair refers to the general mechanism of DNA repair that works by removing the damaged portion of a DNA molecule. During excision repair, bases or nucleotides are removed from the damaged strand. The gap is then patched using complementarity with the remaining strand. This occurs by one of the two mechanisms—base excision repair or nucleotide excision repair.

Base excision repair   Certain mutations are recognized by an enzyme called DNA glycosylases which breaks the N-glycosidic bond between the damaged base and its sugar ring in the DNA backbone. DNA glycosylase can recognize a single, specific type of damaged or inappropriate base in DNA. Certain glycosylases may also introduce a nick 3′ to the site of the damage through associated AP lyase activity. DNA glycosylase acts upon single bases and has no specificity for larger or more complex lesion-involving multiple bases. Interestingly, glycosylases repair or remove the nitrogen base pair by flipping them out of the interior of the double helix in a process called base flipping. DNA glycosylases generate an AP site (a purine/a pyrimidine), i.e., a sugar without nitrogen base. DNA glycosylase has narrow substrate specificity and no cofactor requirement. Once an AP

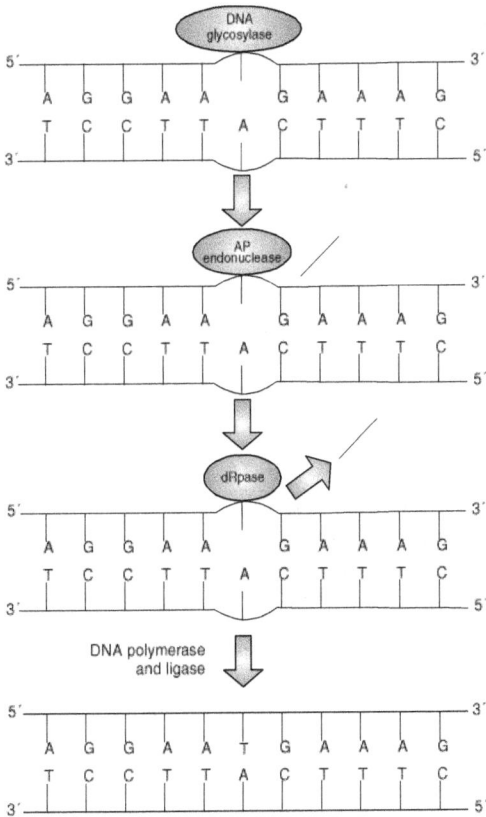

Figure 3.10 Excision repair

site is generated, it is recognized by a second repair-specific enzyme termed as AP endonuclease, which introduces a nick 5′ to the AP site. In *E. coli*, there are two types of AP endonucleases.

1. *Endonuclease* III   This enzyme is encoded by the Xth gene and is important for nicking at 5′ end.

2. *Endonuclease* IV   This enzyme is coded by the *nf* gene and is involved in oxidative damage repair.

Following the 5′ incision to the AP site, a further enzyme deoxyribophospodiesterase (dRpase), is required to hydrolyse the 5′ residue resulting in a single nucleotide gap. DNA polymerase may initiate repair synthesis from this gap, replacing the missing nucleotide residue with a repair patch of one nucleotide. Finally the nick remaining after repair synthesis is closed by DNA ligase.

Nucleotide excision repair   Base excision can repair only one base pair change whereas bulky base damage including thymine dimers are removed by nucleotide excision repair. If the dimerization is not reversed by the photo-reactivation system, then nucleotide excision repair comes into picture. Two copies of the protein product of the uvrA gene combines with one copy of the product of the uvrB gene to form a uvrA2–uvrB complex that moves along the DNA, looking for damage. When the complex finds damage such as a thymine dimer, with moderate to large distortion of the DNA double helix, the uvrA2 dimer dissociates, leaving the uvrB subunit alone. This causes the DNA to bend and attract the protein product of the uvrC gene, uvrC. Binding of uvrC protein causes a conformational change in the DNA following uvrB to nick the DNA four to five nucleotides on the 3′ side of the lesion. After the 3′ nicking by uvrB, uvrC protein nicks 5 bp away, towards the 5′ end. The binding of uvrC and nicking reactions require ATP binding but not hydrolysis. The three components, uvrA, uvrB and uvrC, are together called the ABC exonuclease. The enzyme helicase-II, the product of the uvrD gene, then removes the 12–13 nucleotides along with uvrC. DNA polymerase I fills in the gap and in the process, evicts the uvrB and DNA ligase close to the remaining nick. This is another relatively simple system designed to detect helix distortions and repair them like base excision repair. Nucleotide excision repair is present in all organisms (Figure 3.11).

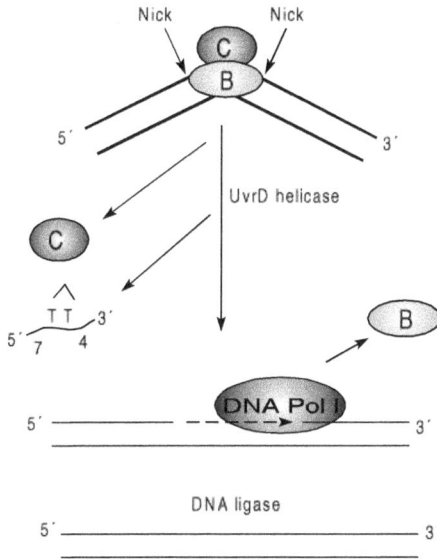

Figure 3.11 Excision repair mechanism

Post-replicative repair When DNA polymerase encounters damage in DNA, it cannot proceed. Instead it gives a gap for replication and proceeds up to 800 bp without replicating. Then again it starts replicating after synthesizing a primer by primosome. These gaps are then repaired by using one of the two mechanisms. Originally several proteins were known to facilitate the replication of DNA with lesions. They were believed to interact with the polymerase to make it capable of using damaged DNA as a template. In *E. coli*, polymerase I can copy damaged DNA. Pol V is error-free and can incorporate ´A´ opposite to thymine dimers. But sometimes, Pol V does errors for unknown reasons, especially during stress. One possible reason for this is that the error-prone polymerase may have developed by evolutionary processes. They create mutations at a time when the cell might need variability. In the second mechanism, a replication fork creates two DNA duplexes. Thus an undamaged copy of the region with the lesion exists on the other daughter duplex. The repair takes place at a gap created by the failure of DNA replication, the process is called post-replicative repair. The RecA protein has two major properties. First, it coats single-stranded DNA and causes the coated single-stranded DNA to invade double-stranded DNA while displacing the other strand of that double helix. RecA continues to move the single-stranded DNA along the double-stranded DNA until a region of homology is found.

The RecA protein is responsible for filling a post-replicative gap in newly replicated DNA with a strand from the undamaged sister duplex. The gap-filling process then completes both strands. The RecA protein is responsible for the damaged single strand invading the sister duplex. Endonuclease activity then frees the double helix containing the thymine dimer. DNA polymerase I and DNA ligase return both daughter helices to the intact state. The thymine dimer still exists, but now its duplex is intact. Another cell cycle is available for photo-reactivation or excision repair to remove the dimer.

SOS-repair   In *E. coli*, it is the last mechanism (till now discovered) by which a cell can repair its damaged DNA. If SOS-repair fails in eukaryotes, then the cell commits suicide by apoptosis. SOS-response is an inducible response to DNA damage resulting in increased capacity for DNA repair, inhibition of cell division and altered metabolism. SOS-response is mediated by the activation of about 20 SOS genes which are normally expressed at low levels due to transcriptional repression by the LexA repressor, which binds to operator sequences (SOS boxes) upstream of each gene. When DNA becomes extensively damaged, single-stranded regions are exposed. Single-stranded DNA interacts with RecA to produce an activated complex termed RecA which facilitates cleavage of the LexA repressor by increasing the rate of LexA autoproteolysis. The cleaved LexA protein can no longer bind to DNA and the SOS genes are de-repressed. When damaged DNA is no longer present in the cell, RecA becomes inactivated and no longer facilitates LexA cleavage. A high level of uncleaved LexA rapidly accumulates in the cell from the existing LexA mRNA pool and the *lex A* gene and other SOS genes are shut down.

Mismatch repair Mismatch or non-Watson–Crick base pairs in a DNA duplex can arise through replication errors, through deamination of 5′ methylcytosine in DNA to yield thymine or thorough recombination between DNA segments that are not completely homologous.

If DNA polymerase introduces an incorrect nucleotide, this error is corrected by mismatch repair (Figure 3.12). Mismatch repair system scans newly replicated DNA or daughter strand and finds the mismatch errors or nucleotides. It uses 3 proteins named Mut S, Mut H and Mut L for the repair. Mut S binds to the mismatch site and it waits there. Then Mut L and Mut H bind creating a loop starting from the site GATC to GATC. This loop formation is driven by the ATP. Then Mut H, which is assumed as endonuclease cleaves at the GATC site on both the sides of the mismatch. Then Mut U unwinds the strand. Then DNA polymerase synthesizes the new strand and ligase joins the strand.

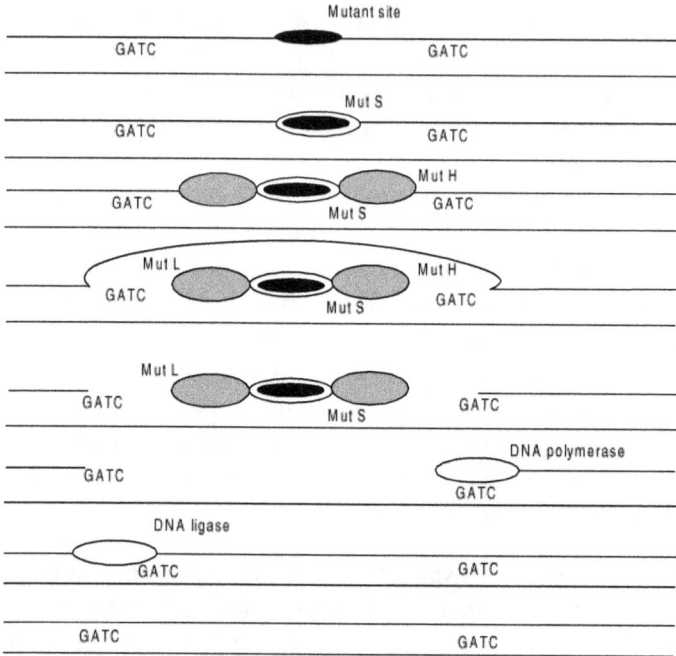

Figure 3.12   Molecular mechanism of mismatch repair

Nucleotide Excision Repair in Eukaryotes

Nucleotide excision repair in eukaryotes involves a large number of genes. The mechanisms of nucleotide excision repair appear to be quite similar in eukaryotes and bacteria, with bimodal incision followed by excision and resynthesis. The repair patch following nucleotide excision repair in humans is slightly longer than that in *E.coli* (30 nt) and is confusingly termed long patch repair, to distinguish it from the 1–2 nucleotide short patch repair which occurs in base excision repair. In eukaryotes, transcription and repair are linked especially for nucleotide excision repair. The basis of this link is the general transcription factor TFIIH, which forms an essential part of both the basal transcriptional apparatus of the cell and the complex of repair proteins, the repairosome. In the transcription complex, the TFIIH core is associated with a cyclin-dependent kinase complex, whereas in repair, it is associated with other repair proteins. Hence, repair is preferentially directed to the transcribed strand of actively transcribed gene (transcription coupled repair) and is more active when transcription is in process (transcription-dependent repair). After DNA polymerase III skips past a lesion such as a thymine dimer, the cell fills in the gap without using template information for complementarity. The site will contain mutations, but at least the integrity of the double helix will be maintained.

Table 3.1 Groups of enzymes involved in different mechanisms

Repair system	Enzymes/Proteins
Base excision	DNA glycosylase
	AP endonuclease
	DNA polymerase
	DNA ligase
Nucleotide excision	UVrA, UVrB,UVrC
	DNA polymerase I
	DNA ligase
Mismatch	Dam methylase
	MutS, MutL, MutH
	Exonuclease
	DNA helicase
	SSB proteins
	DNA polymerase II
	DNA ligase

REVIEW QUESTIONS

1. Describe the three different strategies possible for DNA replication.

2. Describe Meselson–Stahl experiment.

3. Explain the role of SSB in DNA replication.

4. Explain the process of prokaryotic DNA replication in detail.

5. Explain the mechanism of rolling circle replication.

6. Write in details about DNA repair mechanisms that exist in prokaryotes.

7. Write in details about DNA repair mechanisms that exist in prokaryotes.

8. Write short notes on.

    i.   Helicase

    ii. DNA polymerase

    iii. Topoisomerase

    iv. DNA gyrase

    v.   Direct repair

    vi. Excision repair

    vii. Nucleotide excision repair

    viii. SOS repair

    ix. Mismatch repair

4

GENE CONCEPT

Gene is the basic functional unit of life. Just as atoms are the functional units in chemistry, so are genes in biology. The atom combines in "n" number of ways to form the massive, complex structures like mineral ore, etc. Similarly the genes combine in "n" forms to form complex individuals. Just like the atoms combine and exchange electrons from one group to the other to form complex atoms, genes also combine with one another to form complex genes, thus providing a scope for complex gene structure formation.

The concept of gene was first introduced by Mendel and he termed them as hereditary factors or elements. This concept was purely hypothetical and did not carry any experimental evidence. Wilhelm Johansen coined the term "gene" in 1909 to describe heritable factors responsible for the transmission and expression of a given biological character, but without reference to any particular theory of inheritance. But evidence generated from experimental work carried out with higher plants suggest genes as the units of inheritance, controlling one phenotypic trait. Based on this data, Morgan in 1926 proposed the particulate gene theory in which he stated that genes are corpuscular and arranged in a linear order on chromosomes like beads on a string.

According to the pre-1940 "beads-on-a-string" concept, the gene was the basic unit of inheritance defined by three criteria— function, recombination and mutation.

1. The unit of function means that a fragment or unit of genetic material controls the inheritance of one unit character or attribute of phenotype.

2. The unit of structure could be operationally defined in two ways

    i.   As the unit of inheritance not sub-divisible by recombination.

    ii. The smallest unit of genetic material capable of independent mutation.

The classical view was that all three criteria defined the same basic unit of inheritance. But this theory was discarded with advances in the DNA structure. Then Sutton introduced a new gene concept, which was elaborated by Muller. This concept was known as classical gene structure. It states that

1. Genes determine physical as well as physiological characteristics.

2. Genes are present on the chromosome, and on a single chromosome there are many genes.

3. Genes occupy a specific position on a chromosome which is called as locus or loci and genes are arranged in single linear order.

MODERN GENE CONCEPT

According to our current molecular concept, the gene is the functional unit of inheritance coding for one polypeptide chain and operationally defined by the *cis-trans* or complementation test. The other two criteria, recombination and mutation, define the unit of structure, which is equivalent to the single nucleotide pair. Since it clearly does not make sense to call each nucleotide pair a gene, emphasis has been shifted to the original definition of the gene as the unit of function. Thus, Mendel's unit of inheritance co-relates more directly with the gene as a unit of function than with the unit of structure or single nucleotide pair.

GENE STRUCTURE AND ARCHITECTURE

The basic structure of a gene is shown in Figure 4.1. Any DNA sequence which codes for a functional protein does qualify as a gene provided it contains some extra regions with specific function. For genes which encode proteins, a distinction can be made between information translated into polypeptide sequence and untranslated information. In bacteria, the translated region (open reading frame or coding region) is equivalent to the gene and genes are usually separated by short internal non-coding regions. The genes at the extreme ends of the operon are also flanked by a non-coding region which may be termed as the 5′ untranslated region (UTR) or leader sequence and the 3′ UTR or trailer sequence. The 5′ UTR controls ribosomal binding and may facilitate attenuator control. The 3′ region often plays an important role in mRNA stability.

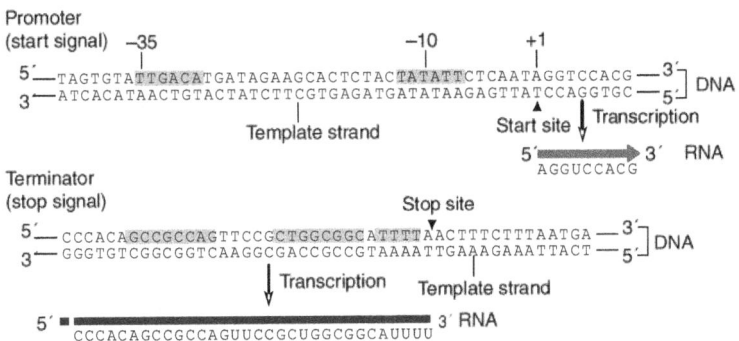

Figure 4.1 Basic structure of gene in whole and its transcription structure

In eukaryotes, the coding region is also flanked by regulatory UTRs. Both UTRs and the open reading frames may be interrupted by non-coding sequences, introns, which are spliced out before RNA exports from the nucleus, i.e., they are not represented in the mature transcript.

Genes are divided into three types based upon the structure.

Simple Gene

A continuous sequence in a nucleotide that codes for a specified polypeptide or functional RNA is called as simple gene. The functional portion of the gene obviously is the sequence of nucleotides that encode for the protein/amino acid that plays a role in the metabolic process of the cell. As RNA sequence is complimentary to the coding strand of DNA, it is customary to represent a gene by complimentary strand sequence of duplex DNA. Transcriptional unit is a sequence of DNA transcribed into a single RNA, starting at the promoter and ending at the terminator. In prokaryotes, a transcriptional unit is continuous but in the case of eukaryotes, we find intergenic spacers. Intergenic spacers are the genomic regions that are not directly involved in the final product and are found between actual coding sectors of two adjacent genes. Spacers are not introns. Spacers are classified into two types based upon their location—leader (present at the 5′ end of the starting of gene) and train (present at the 3′ end of the gene).

Leader region   In relation to the coding strand, the leader sequence will be present at the 3′ end and the trailer will be present at the 5′ end of the coding sequence. In relation to mRNA, the leader sequence will be towards (not present at) the 5′ end and the train present at the 3′ end. Leader sequences are further divided into two types, promoters and enhancers, based upon the position and function they do.

Promoters   Promoters are sequences present in the gene where the transcription enzymes (e.g. RNA pol) and factors (e.g. TBP) bind (Figure 4.2). Promoters help in locating, reading and processing the gene. The promoter defines DNA sequences for initiation reaction for transcription. The maximum size of the promoter is 12 bp, with 90% of purine.

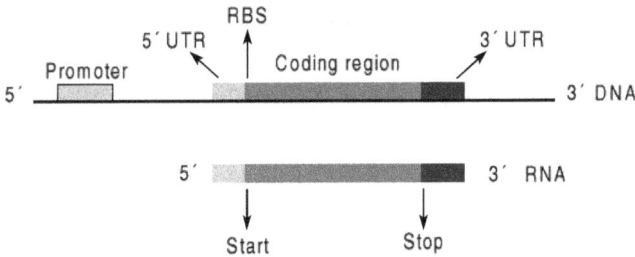

Figure 4.2   Basic structure of gene with promoters

Enhancers   Enhancers are sequences present hundreds of base pairs away from the start point of gene and function as promoters. In contrast to promoter, it is G–C-rich. Enhancers can be present either upstream or downstream from the promoter. Enhancer function is not dependent on the orientation or direction and position, i.e., upstream or downstream, in a given gene. Enhancer can activate any kind of gene to which it is linked. There are no specific enhancers for specific gene types. For example, if an enhancer from a gene is linked to a new gene, still the enhancer will express the new gene without any trouble. They equally activate whether they lie in same or opposite orientation, upstream or downstream and what gene is attached to it.

Train   Trains contain signals for termination of transcription. Usually there exists a continuous sequence of A–T nucleotides, which forms a stem-and-loop structure.

Terminators   RNA polymerase continues transcription until it reaches terminator sequences. At this point the enzyme stops adding nucleotides to the growing RNA chain, releases the transcript and dissociates from the DNA template. Many pro- and eukaryotic terminators form hairpin-like structures at this site thus allowing the delaying of transcript processing. This delay facilitates the binding of rho protein (in some cases) to the RNA polymerase. But at some terminators, termination events can be prevented by specific ancillary sequences (factors) that interact with RNA polymerase. Antitermination causes the enzyme to continue transcription past the terminator sequence, an event called read through. Ancillary proteins are involved in antitermination process.

Compound Gene or Discontinuous or Mosaic or Split Genes

Not all genes are continuous like simple genes, but some genes contain some sequences which do not code for amino acid. The sequence which codes or contains the information coding is called as exon. The sequences which contain the non-coding information are called as introns (Figure 4.3). Introns and exons are DNA sequences, but depending upon whether they are represented in the amino acid sequences or not, they are classified.

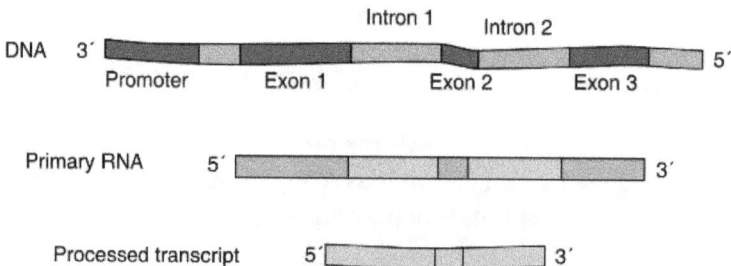

Figure 4.3   An example of intron and exon position in gene

Discontinuous genes or split genes are very common in higher organisms and in some types of bacteria, e.g. archaebacteria. Often the introns are much longer than the exons, for example, human cystic fibrosis when it does not function correctly, is 250 kb from beginning to end and is split into 24 exons and 23 introns. The average length of exon is 227 bp (2.4% of the total length of the gene). These exons are scattered throughout the entire length of the gene, separated by introns of length 2 to 35 kb.

Based upon the number of introns in a split gene, the genes are of two types.

Mono-intron genes These are genes with only one intron, e.g. tRNA *tyr* gene. The transcripts of such genes are not used for translation and they as such perform a biological function.

Multi-intron genes These are genes with more than one intron, e.g. tRNA *tyr* gene. Transcripts of multi-intron genes are translated into proteins, e.g. rat muscle X-acting gene has six introns.

Complex Gene

Immunoglobulin gene, dimorphic and cryptomorphic genes are grouped as complex genes. These genes do not have defined exon or intron sequences, the exon in one instance might be intron in another case. A single complex gene may code for many types of proteins each with affinity to different proteins or function. These genes show excessive rearrangement in sequence (mRNA), i.e., before post-transcriptional modifications. For example, immunoglobulin mRNA is cleaved through a series of steps before it is translated. There is small difference between post-transcriptional modification, i.e., intron splicing and this method. In these genes, in one case, intron 1, 2, 3 are removed whereas in the other case intron 1, 3 are removed and intron 2 acts as exon, thus making an antibody with an affinity towards the new protein or virus.

Pseudogenes

Pseudogenes are genes which have lost their ability to perform a function due to mutations. These genes are very similar to the other members of their family, but contain scrambled nucleotide sequences, e.g. $\psi\varepsilon_b$, $\psi\alpha$ genes look like α and β globin genes.

Overlapping Genes

Most genes do not share information with other genes in the perfect sense. They are discrete, non-overlapping units. In contrast, some genes code for more than one protein or polypeptide; such genes are called as overlapping genes. Overlapping genes share some of the same sequences.

In principle, genes can overlap at two levels. In bacterial systems and other situations where space conservation is necessary, genes may overlap at the level of the reading frame, so that the same information is used to generate two or more unrelated proteins. The open reading frame may be transcribed from the opposite strand, may be translated in different directions and may be out of frame with respect to each other. The genes literally do not have anything in common except that they share the same space. If we look at replicase and coat protein genes of bacteriophage Ms2, these two genes are translated in the opposite direction and in a different reading frame.

Nested Gene

A nested gene is a gene which produces two proteins simultaneously from a long single transcript by changing the end point of protein synthesis.

This can occur by leaky read through of a termination codon or by co-translation frame shifting. F-plasmid, *traX*-gene and eukaryotic RNA viruses use similar strategies. Nested products can also be produced from eukaryotic genes by alternative splicing or by the use of alternative polyadenylation sites.

CIS-TRANS OR COMPLEMENTATION TEST

Complementation test is done to confirm the site of mutation. In complementation test, using a known mutation as reference, an unknown mutation is identified. In this test, two mutations are placed in common protoplasm either is *cis* or *trans* configuration. The *cis-trans* test provides an operational definition of the gene as the unit of function—the unit controlling the synthesis of one polypeptide.

CIS-test   In this method, the two mutations are being examined in a common protoplasm in *cis* or coupling configuration, i.e., both the mutations will be on the same chromosomes. *Cis* test is used to find whether a given mutation is in the mutation site/gene or it is in other site (Figure 4.4). The *cis*-test is an important control test, which establishes the validity of the correlated *trans* or complementation test.

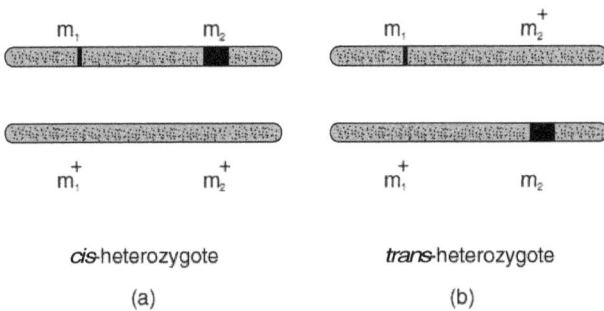

$$ m_1 \qquad m_2 \qquad\qquad m_1 \qquad m_2^+ $$

$$ m_1^+ \qquad m_2^+ \qquad\qquad m_1^+ \qquad m_2 $$

cis-heterozygote   *trans*-heterozygote

(a)   (b)

Figure 4.4   Diagram showing the location of gene in *cis* and *trans* test

Trans test   In this method, the two mutations that are being examined are placed in common protoplasm in *trans* or repulsion configuration, i.e., both the mutations are on different chromosomes.

The concept of *cis–trans* test functions is simple. If an unknown mutation occurs in a known gene mutation, then the test gives a negative result that is there will be no production of product or we will get an inactivated product. If the unknown mutation does not occur in the known gene mutation site, then the test gives a positive result, that is, there will be production of product, because we know that there are two copies of the gene. If one gene is mutated, the other will supplement or code for the protein. As both the mutations are on different genes, the other gene will take the role of coding for the protein.

In the case of a mutation which gives a negative result, there will be no production of protein as both the genes have mutation. In many cases, only the *trans* test is done to confirm the mutation site, since construction of a *cis* heterozygote is time-consuming and expensive. For the *trans* test, one known mutant animal or cell is crossed with unknown mutant animal or cell to get a diploid cell containing the two mutations in the trans or repulsion configuration and is called as trans-heterozygote. The results of the *trans* test are totally uncontroversial, when mutations lead to the partial or total defective protein or total abolished product. This type of situation arises only when mutations are caused by deletions, frameshift mutations or chain termination mutations. The analysis becomes complicated when the mutations are caused by substitution involving different nucleotides, but no alteration in the amino acid charge, or volume of amino acid. The information provided by *cis–trans* or complementation test is totally distinct from that obtained from recombination analyses. Complementation and recombination should not be confused. They are different and give different information. Recombination tells the relationship between two genes whereas complementation gives information about the mutation position.

Even though the *cis–trans* test is a widely used and very helpful technique, it has some drawbacks.

1. It cannot be used when the mutations occur in dominant or co-dominant genes.

2. It cannot be used for polar mutation analysis.

3. It cannot be used for understanding those genes which do not code for diffusible proteins or those which are used in regulatory mechanism.

4. It cannot be used for genes which code for a polypeptide which integrates into a complex protein or which forms dimers.

FINE STRUCTURE OF R-II GENES

The fine structure of a gene is essentially the linkage map of multiple alleles of a gene and it depicts the location of these alleles within the genes. The procedure used for this purpose is basically the same as that used for preparation of linkage maps of different genes. Benzer prepared the first fine structure map in 1961 for the R-II locus of the T4 bacteriophage of *E. coli.*

Bacteriophage T4 is an obligate parasite that grows on *E. coli.* It consists of a single linear molecule of DNA about 200,000 bp long. When a wild type T4 phage particle infects an *E. coli* under optimal conditions, it will produce 200–400 progeny viruses in 20–30 min. by lysing the cell. On a petri plate consisting of lawn of bacteria, wild type T4 phage produces a clear area in the lawn. When present, such an area is called as plaque. These plaques are fuzzy, turbid or halos. When a wild type T4 infection *E. coli* , it is super-infected with a second T4 phage (artificial second infection by strains of the same type of T4 virus). The plaque formation is inhibited by 2 hours. This phenomenon is called as lysis inhibition.

But certain mutations in T4 bacteriophages lead to prevention of lysis inhibition, i.e., the mutants do not show lysis inhibition. Such mutants are called as r-mutants. These r-mutants produce larger plaques. Benzer discovered that r-mutants are due to mutations in one of the three regions or loci, which he called as r I, r II and r III loci. Benzer differentiated these loci due to inability of the r-mutants to grow on different substrains of *E. coli*. r-mutants with mutations in r-II locus or r-II locus mutants are unable to multiply in K12(λ) strain of *E. coli*. But they grow on the other strain of *E. coli* like B and K12. Phages carrying r-II mutants can be easily isolated by serially transferring inoculum with sterile loop from r-type plaques on *E. coli* B lawn to lawns of *E. coli* K12 (λ), B and K12. The r-I and r-III mutants will grow on both K12 (λ) and B whereas r-II will grow only on B lawn but do not grow on K12(A).

To confirm the mutation site, Benzer did *cis–trans* or complementation tests, and all the r-II mutants were found to be located in one of the two genes. For the complementation tests, Benzer developed a simplified procedure called complementation spot test. In this method, the mutant phage which needs to be tested is mixed with *E. coli* K12(λ) strain and r^+ mutant grows on K12(λ) strain. Then he plated the mixture on the nutrient medium. Then a large plaque is formed at the reciprocative site, i.e., if the mutation has occurred in r-III B gene, then the plaque is formed at the drop placed in r-II A gene. If complementation does not occur, the area will be grown with the lawn of *E. coli* cells. r^+ mutants are mutants that occur at other than the three sites (r-I, r-II and r-III).

Although, complementation tests showed that all the r-II mutations were located in two genes, recombination tests demonstrated that these mutations were located at many distinct sites within these two genes. Recombination between r-II mutations is analysed by simultaneously infecting *E. coli* B cells (permissive host cells). With the two r-II mutants in question and plating the progeny phages on a lawn of *E. coli* K12(λ) will show if any wild type recombinants have been produced. The total number of progeny phages produced is determined by plating the lysate on an *E. coli* s lawn. As many as 10^8 progeny phages may be seen in a single petri plate as recombination is a very rare event (2 in 100 million and can be easily be detected.)

Deletion Mapping

Many of the spontaneous r-II mutants isolated by Benzer failed to recombine with two or more mutants that recombine with each other. They behaved as though they contained "multisite" mutations or defects that extended over a segment of the r-II locus. Some of them appeared to span the entire rII, others only a short segment. Benzer proposed that these multisite mutations resulted from the deletion or loss of segments of the DNA. As they failed to undergo back mutation, Benzer´s proposal has subsequently been verified directly by analysis of several deletion mutations by electron microscopy.

Electron Microscope Heteroduplex Mapping

The existence of genetically well-defined deletion mutations at the r-II locus has permitted E.K.F Bautz and colleagues to determine the physical size of the r-II locus. This was done using a technique called heteroduplex mapping.

A heteroduplex is a DNA molecule in which the strands are not entirely complementary. Heteroduplex mapping involves the *in vitro* preparation of DNA heteroduplexes and their analysis by electron microscopy. The wild-type T4 phage DNA is isolated after growing in *E. coli* cells with radiolabelled PO_4. The mutant phage was also grown on *E. coli* cells without radiolabelled PO_4. Then DNA is isolated from both the wild and mutant phages and is denatured by heating to 100°C. The denatured DNAs are then mixed together and then allowed to renature by incubating the mixture at 60°C for 12 hours. Some heteroduplexes will be forming with one strand of DNA from the wild type T4 and one strand from the mutant phage. The wild type strand of these heteroduplexes will not have complementary base sequences with which to pair in the segment of the mutant type DNA molecule. The wild type strands will thus form single-stranded loops, the length of which can be measured on electron micrographs.

Bautz and co-workers prepared heteroduplexes between DNA from T4 r⁺ phage and DNA from each of several genetically well-characterized r-II deletion mutants; they then analysed them by electron microscopy. Their results yielded estimates of 1800 ± 70 bp for r-IIA and 845 ± 50 bp for r-IIB locus. These results combined with the extensive genetic data of Benzer and colleagues, provide a fairly clear picture of the fine structure of the r-II locus.

General Conclusions

1. The various r-II mutants mapped at about 400 different sites in cistrons A and B.

2. Both muton and recon represent one base pair.

3. The frequency of mutation at different sites is quite variable.

4. Some sites showed very high frequency of mutation; these are called hot spots.

REVIEW QUESTIONS

1. Explain beads on string.

2. Write the salient features of modern gene concept.

3. Define the gene structure.

4. Explain the architecture of simple gene.

5. Write in detail about compound gene.

6. Explain what is complement test.

7. Define r-II locus.

8. What is deletion mapping?

9. Write short notes on the following.

    i.   Leader region

    ii.   UTR

    iii. Promoter gene

    iv.   Complex gene

    v.   Nested gene

    vi.   Overlapping gene

5

TRANSCRIPTION OR GENE EXPRESSION

The presence of a gene in the genome is not sufficient. It should express or it should make protein, only then does it have some significance or use for the organism. Genes make proteins by a process called gene expression. Gene expression takes place in two steps or phases.

1. Transcription or RNA synthesis

2. Translation or protein synthesis

Transcription is the first step in gene expression and the predominant level of gene regulation. Transcription is a process in which one type of nucleic acid produces another type of nucleic acid or it is the process of RNA synthesis from DNA molecule with the help of complex protein. The concept of mRNA synthesis is to carry information from a gene located in the nucleus to the ribosome in the cytoplasm. This was first given by Crick in 1958. Transcription and translation are depicted in Figure 5.1.

The following are the points of similarity between DNA and RNA synthesis.

1. The general steps of initiation, elongation and termination with 5′–3′ polarity.

2. Large, multi-component initiation complex.

3. Adherence to Watson and Crick base-pairing rules.

RNA synthesis differs from that of DNA in the following.

1. Ribonucleotides are used instead of DNA.

2. T is replaced with U.

3. A primer is not involved.

4. Only a small portion of DNA is used.

5. No proofreading activity.

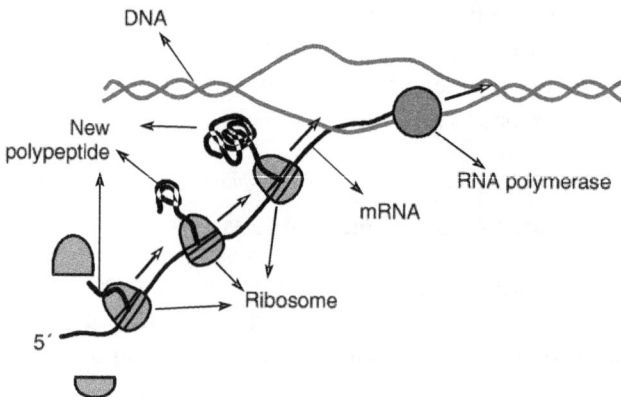

Figure 5.1 Coupled transcription and translation

The sequence of ribonucleotides in an RNA molecule is complementary to the sequence of deoxyribonucleotides in one strand of the double-stranded DNA molecule. The strand that is transcribed into an RNA molecule is referred to as the template strand of the DNA. The other DNA strand is frequently referred to as the coding strand.

TRANSCRIPTION IN PROKARYOTES

Transcription is much simpler than translation and is accomplished by a single enzyme in conjugation with a variety of auxiliary proteins. The enzyme responsible for transcription is called as DNA-dependent RNA polymerase or simply RNA polymerase. In prokaryotes, e.g. *E. coli*, RNA polymerase enzyme is a holoenzyme, i.e., it is composed of a core enzyme and a protein called sigma factor. The core enzyme is made up of four subunits [two copies of α (41,000 Dalton each) β (155,000 Daltons) and β' (165,000 Daltons)]. The exact function of these subunits is clear. The β subunit is responsible for binding of ribonucleotides prior to RNA synthesis and β' is responsible for attachment to the DNA template. The function of the core enzyme is the polymerization or catalysis of covalent chain extension. But the core enzyme cannot start the transcription by itself, it needs the help of another protein called sigma factor. Sigma factor is a 95,000-dalton protein that helps in recognition and binding of core enzyme to initiation sites or promoters on DNA. Not only that, it converts the closed promoter complex into an open promoter complex. The sigma factors recognize both the –35 and –10 sequences. The sigma factor is also sensitive to the spacing between these sequences. The use of different sigma factors is one mechanism the cell uses to select a particular battery of genes for possible transcription.

RNA polymerase	Core enzyme + Sigma factor
Core enzyme	BB'
Holoenzyme	Core enzyme + Sigma factor or simply

RNA Polymerase

RNA polymerase is a 350-Kdalton core complex. RNA Pol is a metallo-enzyme containing 2 zinc molecules. The core RNA polymerase associates with a specific protein factor (sigma factor) that helps the core enzyme to attach more tightly to promoter.

Transcription of nongenic regions or genes coding for enzymes not necessary is wasteful. Hence, RNA polymerase should be selective. It should use as transcription templates only those DNA segments whose products the cell needs at that particular time. RNA polymerase must be able to recognize both the beginnings and the ends of genes on the DNA double helix in order to initiate and terminate transcription. It must also be able to recognize the correct DNA strand to avoid transcribing the DNA strand that is not informative. RNA polymerase accomplishes these tasks by recognizing certain start and stop signals in DNA, called initiation or promoter termination sequences. The DNA region to which RNA polymerase molecule binds prior to initiating transcription is called the promoter. The promoter is an important part of the gene as it contains information for transcription initiation sites and the information that determines which of the two DNA strands is transcribed (Figure 5.2). Promoters are more than simply recognition regions, they can act as important control sites in regulating the rate of gene expression. Mutations that occur in the promoter and tend to make the promoter weaker, resulting in less transcription, are called as down mutations. And the reverse where mutations make a promoter stronger are called up mutations.

Promoters are located before the starting point of a gene and they are mostly conserved with very slight modifications. There are 3 types of promoters that are widely found in most genes.

Pribnow box The conserved sequence (TATAAT) present just 10 base pairs before the first transcribed base is called as pribnow box. This region is 6 nucleotides in length. It is also called as TATA box or open complex, as it is the site from where the actual unwinding of DNA starts. The nucleotides in the TATA box are mostly adenine and thymine, hence the DNA denaturation occurs (during transcription) very easily. The polymerization reaction starts six to eight nucleotides down from the pribnow box.

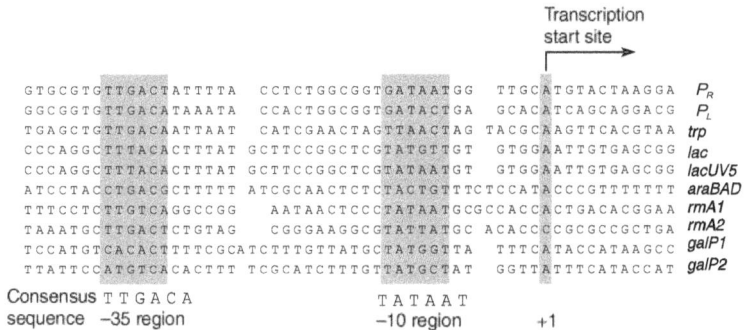

		Transcription start site	

```
GTGCGTGTTGACTATTTTA   CCTCTGGCGGTGATAATGG   TTGCATGTACTAAGGA   PR
GGCGGTGTTGACATAAATA   CCACTGGCGGTGATACTGA   GCACATCAGCAGGACG   PL
TGAGCTGTTGACAATTAAT   CATCGAACTAGTTAACTAG   TACGCAAGTTCACGTAA   trp
CCCAGGCTTTACACTTTAT   GCTTCCGGCTCGTATGTTGT   GTGGAATTGTGAGCGG   lac
CCCAGGCTTTACACTTTAT   GCTTCCGGCTCGTATAATGT   GTGGAATTGTGAGCGG   lacUV5
ATCCTACCTGACGCTTTTT   ATCGCAACTCTCTACTGTTTCTCCATACCCGTTTTTT   araBAD
TTTCCTCTTGTCAGGCCGG   AATAACTCCCTATAATGCGCCACCACTGACACGGAA   rmA1
TAAATGCTTGACTCTGTAG   CGGGAAGGCGTATTATGC  ACACCCCGCGCCGCTGA   rmA2
TCCATGTCACACTTTTCGCATCTTTGTTATGCTATGGTTA   TTTCATACCATAAGCC   galP1
TTATTCCATGTCACACTTT  TCGCATCTTTGTTATGCTAT   GGTTATTTCATACCAT   galP2

Consensus  T T G A C A                    T A T A A T
sequence   –35 region                    –10 region        +1
```

Figure 5.2 Different promoter sequences in different genes

CAT box Another conserved sequence–TGTTGACA–present 35 base pairs before the first transcribed base is called as CAT box or closed complex. This complex is eight nucleotides in length.

Upstream element These are A.T-rich sequences present between 60 to 40 base pairs before the first transcribed base. These are strong bacterial promoters and control how a gene is expressed and to what level. The x-protein in core enzyme recognizes these upstream elements.

Terminators Transcription continues as RNA polymerase and nucleotides are added to the growing RNA strand according to the rules of complementarity. The polymerase moves down the DNA until the RNA polymerase reaches a stop signal or terminator sequence. Two types of terminator mechanisms, rho-dependent and rho-independent, differ in their dependency on the rho protein.

BASIC MECHANISM OF TRANSCRIPTION

Like other polymerization reactions, transcription is divided into three stages or steps (Figure 5.3).

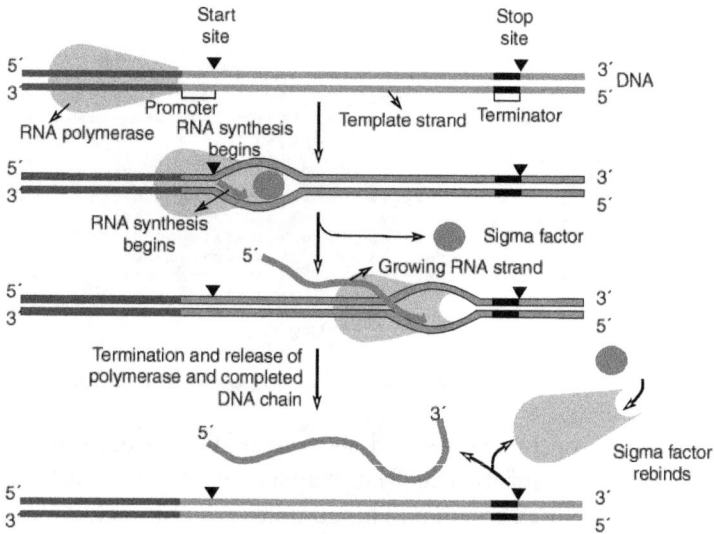

Figure 5.3 Simple diagram of transcription

Initiation

First, the RNA polymerase enzyme recognizes the promoter regions, which lie just "upstream" of the gene and bind to it. This initial structure formed between RNA polymerase and DNA is called the closed promoter complex. In this complex the enzyme covers or protects about 60 bp of the double helix, starting from just upstream of the –35 box to just downstream of the –10 box. Then in –10 box the DNA helix unwinds due to breakage of base pairing (called melting). Melting is an essential prerequisite for transcription because the bases of the template strand must be exposed in order to direct transcript synthesis. This open structure is called as its open promoter complex.

When the open promoter complex is forced, the first two ribonucleotides are base paired to the template polynucleotides at positions +1 and +2 and the first phosphodiester bond of the RNA molecule is synthesized by RNA pol. Transcription follows the same base pairing rules as DNA replication; T, G, C and A in the DNA pair with A, C, G and U respectively. This base pair pattern ensures that an RNA transcript is a faithful copy of the gene. The end of initiation is signified by promoter clearance, where the RNA polymerase moves away from the promoter site without dissociating, freeing the promoter from further initiation events. Promoter clearance occurs only if the open promoter complex is stable and usually follows a number of abortive initiations where short transcripts are generated. Initiation is usually the rate-limiting step in transcription and is the primary level of gene regulation in both prokaryotes and eukaryotes.

Elongation

During the elongation phase of transcription, the RNA polymerase moves along the template synthesizing nascent RNA in the 5′–3′ direction. The double-stranded DNA is unwound ahead of the elongation complex and rewound behind it. The dynamic, transiently melted structures representing the site of transcription is called as "transcription bubble". The transcription bubble is between 12 to 17 base pairs in length and is constant throughout transcription. Thus RNA polymerase has unwindase and windase activity as no other enzymes such as helicase or topoisomerase are taking part in the transcription. The underlying chemical reaction in transcription elongation is synthesis of an RNA molecule. During polymerization the 3′-OH group of one ribonucleotide reacts with the 5′-PO_4 of the second ribonucleotide to form a phosphodiester bond. This results in loss of a pyrophosphate molecule (Pi) for each bond formed. In transcription, the chemical reaction is modulated by the presence of DNA template, which directs the order in which the individual ribonucleotides are polymerized into the RNA.

As the polymerase moves along, it must insert the proper nucleotide into the growing chains at each site. Presumably the enzyme is capable of selecting the complementary ribonucleotide triphosphate for incorporation as a result of the ability of the nucleotide to form the proper stereochemical fit with the nucleotide in the DNA strand being transcribed. A bacterial RNA polymerase is capable of incorporating approximately 50 nucleotides/second into the growing RNA molecule.

The new RNA is designated as transcript. The RNA transcript is therefore synthesized in the 5′–3′ direction and is built in a step-by-step fashion with new ribonucleotide being added to the free 3′ end of the existing polymer. The transcript is longer than the gene. Rarely, position +1, where synthesis of the RNA transcript actually begins, will be the start point of the gene itself. It is almost invariably the RNA polymerase enzyme that transcribes the leader segment before reaching the gene. Leader segment is the sequence present before the first codon (AUG), the length of this leader varies from gene to gene and it may be as short as 20 nucleotides, or longer than 600 nucleotides. Similarly when the end of the gene is reached, the enzyme continues to transcribe a similar segment before termination occurs. The elongation rate was usually constant although it may be perturbed by secondary structure in the template.

Termination

As with initiation, termination is not a random process but must occur only at a suitable position shortly after the end of all genes. The polymerase moves down to the DNA until RNA polymers reached a stop signal or terminator sequence. Termination can occur either with the help of hexamer protein called rho or without it. Termination that occurs with the help of rho protein is called as rho-dependent termination and that which occurs without the help of rho protein is called a rho-independent termination. Interestingly, both types of terminators sequenced so far have a common sequence capable of forming an inverted, repeat sequence. The repeated inverted sequences are separated by four base pair sequence. Inverted repeats can form a stem-loop structure by pairing complementary bases, within the transcribed messenger RNA. In fact, the inverted repeat can exert its influence on transcription by enabling a stem-loop to form in the growing RNA molecule (Figure 5.4). A cruciform structure is

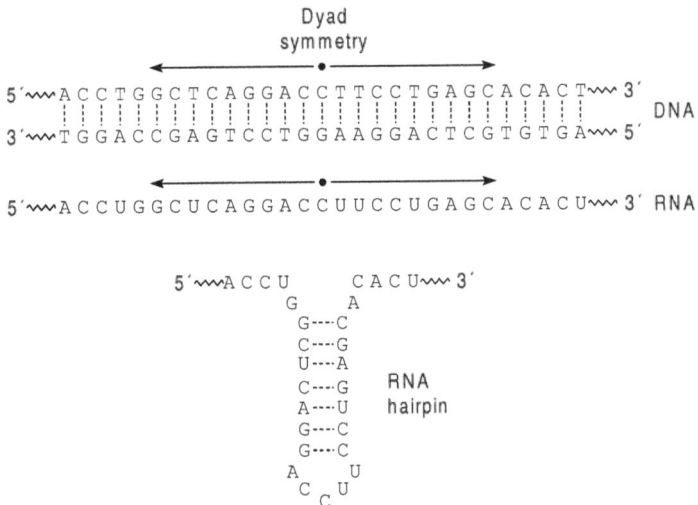

Figure 5.4 Molecular structure of the RNA hairpin formed during transcription termination step

unlikely to form in the double-stranded DNA because the normal double helix is more energetically stable due to the much larger number of potential base pairs that can form. Remember that, only 17 base pairs are in single-stranded form.

Both rho-dependent and rho-independent terminations involve the formation of stem-loop. It appears that the RNA stem-loop structure formed causes the RNA polymerase to pause just after completing it. This pause may then allow termination under two different circumstances. In rho-independent terminators, the pause may occur just after the sequence of uracil is transcribed. Perhaps during the pause the uracil–adenine base pairs spontaneously denature, releasing the transcribed RNA and the RNA polymerase terminating the process and making the polymerase available for further transcription.

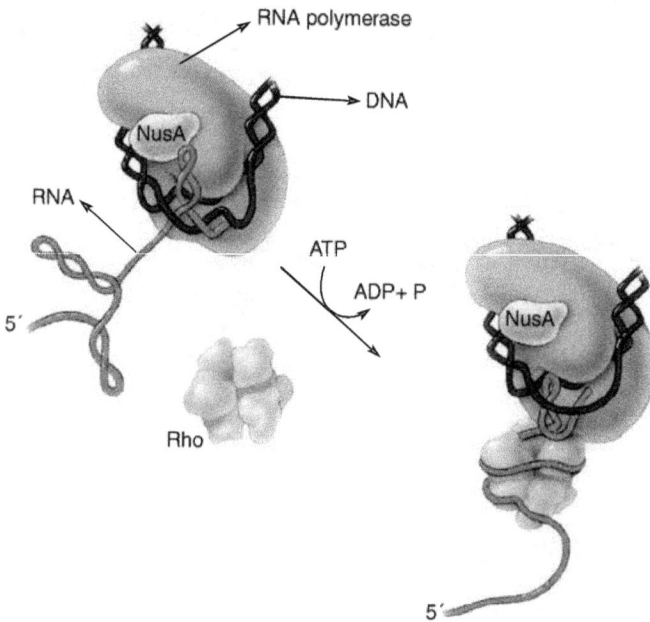

Figure 5.5   Molecular structure formed during rho-dependent transcription termination

Rho-dependent terminators do not have the uracil sequence after the stem-loop structure. Here, termination depends on the action of rho, which appears to bind to the newly forming RNA (Figure 5.5). In an ATP-dependent process, rho travels along the RNA at a speed comparable to the transcription process itself. Possibly, when RNA polymerase pauses at the stem-loop structure, rho catches up to the polymerase and unwinds the DNA–RNA hybrid. Rho can do it as it has DNA–RNA helicase properties. Yet, the exact sequence of events at the terminator is not fully known presently.

TRANSCRIPTION IN EUKARYOTES

Transcription in eukaryotes is a very complex process involving one of the groups of RNA polymerase enzyme and a number of associated proteins. In eukaryotes, different RNA polymerases exist for different types of genes. The process of eukaryotic transcription is less understood and certain RNAs, mRNAs in particular, have very different lifespans in a cell, thus complicating the situation. Not only this, the RNA molecules synthesized in mammalian cells are very different from those made in prokaryotes.

The various components of eukaryotic transcription are as follows:

RNA Polymerases

There are three different sets of RNA polymerases. RNA Pol I is found in the nucleolus and is responsible for transcription of genes for rRNA. RNA Pol II is responsible for synthesis of heterogeneous nuclear RNA or mRNA. RNA Pol III is responsible for transcription of small nuclear RNAs and tRNAs. The sizes of the RNA polymerases for the three major classes of eukaryotes range from molecular weight 500,000 to 600,000. The enzymes are complex with more than 14 subunits which assist or serve as regulators. The functions of each of the subunits are not yet understood. The eukaryotic RNA polymerase has extensive amino acid homologies with prokaryotic RNA polymerase.

Surprisingly, all eukaryotic RNA polymerase have similar structure and to some extent same amino acid sequence, but they respond to one peptide toxin, (α-amanitin) in different ways. α-amanitin is thought to block the translocation of RNA polymerase during transcription. RNA Pol I is insensitive to α-amanitin, but RNA Pol II is sensitive even at low concentration whereas RNA Pol III is sensitive at higher concentrations of α-amanitin. Eukaryotic Pol II consists of 14 subunits. The two largest subunits about 200 kD, are homologous to the bacterial β and β' subunits. Eukaryotic Pol II has a series of heptad repeats of sequence Tyr-Ser-Pro-Thr-Ser-Pro-Ser at the carboxyl terminal of the largest Pol II subunit. This carboxyl terminal repeats domain (CTD) has 26 repeated units in yeast and 52 units in mammalian cells. The CTD is both a substrate for several kinases, including the holoenzyme component TFIIH, and a binding site for a wide array of proteins known as srb or mediator protein. The CTD has also been shown to interact with RNA processing enzymes. Pol II is activated when phosphorylated on the Ser and Thr residues and is inactive when the CTD is dephosphorylated. RNA Pol II can start transcription only when it has CTD. Srb proteins or mediators bind to CTD. The exact role of these bindings is not clear yet.

Promoters

Promoters of eukaryotes are somewhat similar to prokaryotes. They are regions where RNA polymerase binds and begin the transcription. In eukaryotes, promoters not only define where transcription should start, they also determine how frequently this event is to occur since there are three different RNA polymerases in eukaryotic nuclei. There are more than three different promoters which are recognized by different polymerases. RNA polymerase does not recognize eukaryotic promoters by itself. For the recognition, RNA polymerase requires the help of or relies on proteins called transcription factors.

Promoters for RNA Pol II RNA polymerase II promoters bear the closest resemblance to prokaryotic promoters. Moreover, polymerase II is responsible for transcribing the vast majority of genes coding for proteins. These promoters confer fidelity and frequency of initiation and have rigid requirements for both position and orientation. Single base changes have dramatic effects on function.

TATA box TATA box or Hogness box at about 30 bp from the start site of transcription, is a conserved A–T-rich sequence (TATAAA). This sequence is also called as TATA box. This sequence has close similarity to that of prokaryote –10 sequences but the difference is that they are located at –30 bp in eukaryotes.

CAT box Another conserved sequence is usually found at about –75 bp from the start site of transcription. It has a consensus sequence GGCCAATCT.

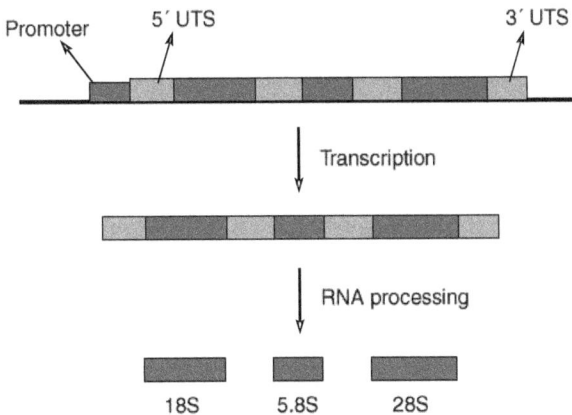

Figure 5.6 Gene structure and expression of rRNA genes

Promoters for RNA Pol III   The polymerase III promoters that govern the 5S transcription of 5S rRNA and tRNA genes transcription are radically different from those recognized by bacterial polymerase. Polymerase III promoters are located within the genes they control. These promoters are present between +50 to +83 bp from the start point of transcription. These promoters are called as internal control region (ICR). 5S RNA gene promoter is actually split into two major elements called box A and box C with a short sequence in the middle called intermediate element. In the case of tRNA, the ICR is split into two major elements box A and box C, without any intermediate element. The sequence between the two boxes is of least importance. Quite strongly, RNA polymerase III transcribes genes encoding other small RNAs which contain promoters not within the gene, but at the 5′-end of the gene starting. One small RNA called 7SL RNA contains one TATA box.

Promoters for RNA Pol I   Promoters of RNA Pol I are like promoters of RNA Pol II, in position only. RNA Pol I promoter base sequences do not match in sequences either to TATA box or CAT box, RNA Pol I recognizes promoters present on rRNA genes. Promoters of this gene are present at the 5′-end of the gene. The first promoter called core element is found between position −45 and +20. This promoter sequence is not conserved like TATA box and is 65 bp in length. The other promoter found between −156 to −107 before the transcription site is called as upstream control element (UCE). This promoter sequence is also not conserved like CAT box and is 49 bp in length.

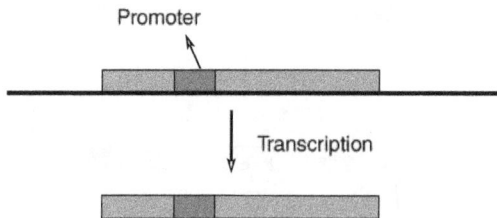

Figure 5.7   Promoter position in the tRNA gene

Transcription Factors

Transcription factors are groups of proteins which assist the RNA polymerase in recognizing promoters. Transcription factors are divided into two types—if they can activate specific genes they are called as gene-specific transcription factors or gene activators, and those which can activate any gene are called as general transcription factors.

The general transcription factors bind to DNA regions within promoters by themselves and deliver the RNA polymerase to their respective promoter sites. General transcription occurs at a low or basal level. But gene-specific transcription can act both positively (boost the transcription rate) or negatively (inhibit transcription).

As different RNA polymerases recognize different promoters, similarly each polymerase is activated by a different transcription factor.

RNA polymerase I transcription factors   The pre-initiation complex that forms at rRNA promoters is very simple and involves the binding of polymerase I in addition to two transcription factors called SL 1 and upstream binding factor (UBF) to the promoter. UBF binds to the upstream control element (UCE) of the rRNA promoter and causes polymerase I to bind near the transcription start site. One UBF binds, SL 1 can join the complex strengthening Pol I binding and thus promote formation of the pre-initiation complex. SL 1 is a complex protein with a peptide called TBP (It is one of the proteins in TATA binding protein (TBP) family), which is similar to TFIID and 3 peptides, which bind to TBP. These three peptides are called as TAFs (TBP-associated protein). These are not similar to TAFs in TFIID.

RNA polymerase III transcription factors   As this polymerase transcribes different types of genes, the transcription factors complex are also different. That is, the transcription factors are same, but the complex formed is different. Three factors are involved in 5S rRNA transcription—TFIIIA, TFIIIB and TFIIIC. TFIIIB is the true transcription factor, while the others are just assembly factors. TFIIIB helps RNA polymerase III bind just downstream, in a position to start transcribing the gene. First TFIIIB and TFIIIC bind to the internal promoter followed by TFIIIB. TFIIIB positions RNA polymerase III at the transcription start site.

In case of tRNA gene transcription only two factors are involved—TFIIIB and TFIIIC. First TFIIC binds to the internal promoter and then TFIIIB binds to TFIIC by protein–protein interaction. TFIIIB contains TBP like S 1 and plays the same role as that of SL-1, whereas TBIIIA and TBIIC play the same role as that of UBF.

RNA polymerase II transcription factors   Transcription factors for RNA Pol II are complex, and interact with many factors before they initiate the transcription. There are thus three classes of transcription factors involved in the regulation of transcription by RNA Pol II. The first class is basal factors, e.g. TBP, TFIIA, etc. Basal factors help in basal transcription. Basal transcription requires a number of factors called TFIIA, B, D, E, F and H. Of all those factors, TFIID is the most important and widely studied transcription factor. TFIID is a complex protein consisting of TBP (TATA box binding protein) and eight TAFs (TBP-associated proteins). TFIID is a 700 kD protein that covers a 35 bp region of DNA. TBP is a 20–40 kD protein that binds to a 10 bp segment of DNA over TATA box. TBP supports only the basal transcription, on the other hand TFIID can support enhanced and basal transcription. There are other transcription factors whose functions are known but other finer details such as what is their molecular weight, who governs their action, etc. are still unclear.

The second class is co-activators, e.g TAFs. TAFs are essential for enhanced transcription. Hence they are also called as co-activators. Different combinations of TAFs with TBP may lead to binding to different promoters and activating them with different strengths. The particular assembly of TAFs at the basal promoter therefore dictates how the initiation complex interacts with upstream activators, enabling the basal promoter to demonstrate some cell-type specificity. There are some factors which enhance the transcription rate converting basal level transcription to enhanced level. Such factors are called as activators.

A third class of sequence elements that can either increase or decrease the rate of transcription initiation are the enhancers or repressors. Enhancers that suppress the transcription are called as silencers or repressors. Enhancers and silencers are same in structure and architecture, expect in the function. Enhancers are DNA sequences present hundreds of base pairs away from the starting point of the gene and enhance the transcription. They stimulate transcription from promoters. They are different from promoters in the sense that they are not relatively fixed in spatial relationship with the genes they control. Enhancers can be present both upstream and downstream from the transcription start site. Enhancers can function in any orientation. Transcription factors can bind to enhancer-like promoters. However, because of their distance, such interactions are mediated by the looping out of the intervening DNA, allowing enhancers to act in a manner independent position and orientation. Interestingly enhancers do not have any conserved sequence like promoters. Enhancers have been found to be associated with a variety of genes. They are usually tissue-specific, that is, a gene with a given enhancer can function only in certain kinds of cells. This tissue specificity derives from the fact that enhancers depend on transcription factors and these factors are made in some tissue but not in others.

Mechanism of Transcription by RNA Polymerase II

A complex system consisting of as many as 60 unique proteins provides accurate and regulatable transcription of eukaryotic genes. RNA polymerase must recognize a specific site in the promoter in order to initiate transcription at the proper nucleotide. In contrast to the situation in prokaryotes, eukaryotic RNA polymerase alone is not able to discriminate between promoter sequence and other regions of DNA. Thus, other proteins facilitate promoter-specific binding of these enzymes and formation of the pre-initiation complex. Another set of proteins, co-activators, help to regulate the rate of transcription initiation by interacting with transcription activators that bind to upstream DNA elements. Like prokaryotic transcription, eukaryotic transcription is divided into 3 steps.

Initiation Unlike prokaryotic transcription, where the binding of sigma factor to RNA Pol initiates the transcription, the situation is more complex in eukaryotic genes. The formation of basal transcription complex is the first step of initiation. The first step in basal transcription complex formation is the recognition of the TATA box by the TFIID factor (consists of TBP and TAFs). TBP makes sequence-specific contacts with DNA at the TATA box through the minor groove instead of using the major groove. Most transcription factors bind to the major groove. The role of TAFs is to form an assembly that directs the formed pre-initiation complex to the promoter. They do not take part or assist in the pre-initiation complex.

Further multimeric factors then bind sequentially to the promoter. TFIIA is the first to be recruited after TFIID. TFIIA blocks the binding of DR1, which inhibits TFIID activity. Then TFIIB binds and acts as bridge protein for the recruitment of TFIIF. TFIIF carries the RNA polymerase II enzyme into the complex. TFIIB and TFIIF may together promote specific interactions between RNA polymerase II and the start site. The enzymes facilitate the binding of TFIIE, which in turn allows binding of TFIIH and TFIIJ. TFIIH is essential in the pre-initiation complex as it has helicase activity, which controls promoter melting and kinase activity which phosphorylates the CTD of RNA Pol II. The addition of TFII E and TFIIM is the final step in the assembly of this pre-initiation complex or basal transcription complex. Each binding event extends the size of the complex so that finally 60 bp are covered.

Binding of TFIIH releases the enzyme from the initial complex and facilitates promoter clearance, leaving the initiation complex at the promoter like bacterial RNA polymerase. Eukaryotic RNA polymerase may make abortive initiations before promoter clearance. The first nucleotide inserted is usually a purine, which is modified immediately by the enzyme mRNA guanyl transferase to generate a distinctive cap.

Elongation   The process of elongation of both eukaryotes and prokaryotes is the same. With minor additions, eukaryotic elongation is not a smooth process; it can encounter secondary structures that lead to the arrest or termination of transcription. To prevent this, various elongation factors are associated with RNA polymerase during elongation. There is evidence that suggests that RNA processing occurs along with elongation site side by side. The elongation step in eukaryotes is slower when compared to bacteria. The transcription is longer due to the presence of long genes. In the case of human dystrophin gene, the largest gene known till date, transcription of 2.5 megabase pairs takes 16 hours to complete.

Termination   The termination of transcription in eukaryotes is poorly characterized. As most RNA polymerase II transcripts are processed by polyadenylation at 3-OH end, the intrinsic termination is not clear.

Mechanism of Transcription by RNA Pol I and Pol III

The transcription of rRNA genes by RNA Pol I and 5S rRNA and tRNA genes by RNA Pol II is similar to that by RNA Pol II, except that there are minor differences in transcription factors which initiate the pre-initiation complex formation. The elongation step is the same. But termination is different from prokaryotic termination and eukaryotic RNA Pol II transcription termination. Termination of RNA polymerase I transcription occurs at a site approximately 15 bp to the end of the mature rRNA and involves recognition of a specific *cis*-acting element. Termination of RNA polymerase III transcription occurs at sites similar to bacterial rho-independent terminators.

POST-TRANSCRIPTIONAL MODIFICATION

In eukaryotes, transcription and translation cannot occur simultaneously because both these processes occur at two different sites. Eukaryotic transcription results in a primary transcript. Three major changes occur in the primary transcript before it is transported into the cytoplasm for translation. These changes include.

1. Cleavage of introns
2. Addition of T-methyl guanosine groups or mRNA capping
3. Polyadenylation or poly(A) tail

mRNA capping

This process is the first of the processing reactions, at least for hnRNA. The cap is a 6-methyl guanosine attached backward through a triphosphate linkage to the 5′-terminal end of the mRNA (Figure 5.8). The nuclear enzyme called guanyl transferase or mRNA guanyl transferase catalyses the addition of the guanosine triphosphate part of the cap. It may be associated with RNA polymerase initiation complex. Another enzyme called guanine 7-methyl transferase present in the cytosol methylates this guanine. The methyl group comes from S-adenosyl methionine and the methyl group at the 7-position in guanosine. This type of capping is called as type zero cap, e.g. yeast. In higher eukaryotes, a further methyl group is transferred to position O^2 of the ribose sugar in the next nucleotide, i.e., it is the first nucleotide in the transcript corresponding to this position to generate a type 1 cap, e.g. human. In this case, two methyl groups will be present one on the guanosine and the other is on the sugar of the other, next or first (+1) residue. If the first residue (+1) after guanosine is adenine, then the methyl group is added at N6 position of the adenine or at O^2 of the sugar. Such a type of capping is called as type 2 Cap.

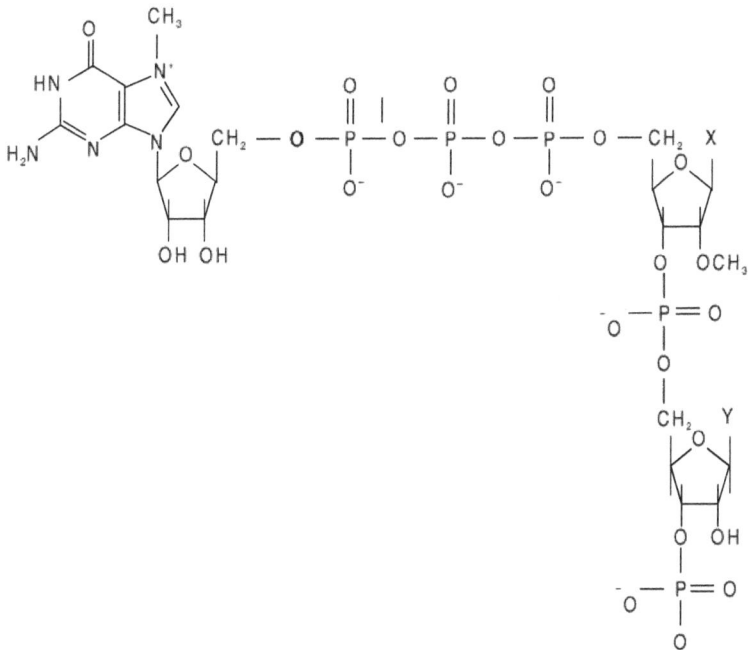

Figure 5.8 mRNA cap structure

The process of RNA capping starts with removal of phosphate group from triphosphate group at 5′-end, leaving the diphosphate group. Then, before the RNA has grown to 50 bases long, a capping enzyme adds GTP to its 5′ end. This reaction is unlike the normal RNA polymerase reaction in several aspects. The incoming nucleotide is added to the 5′-end of the RNA chain instead of the 3′-end. Instead of a simple phosphodiester bond

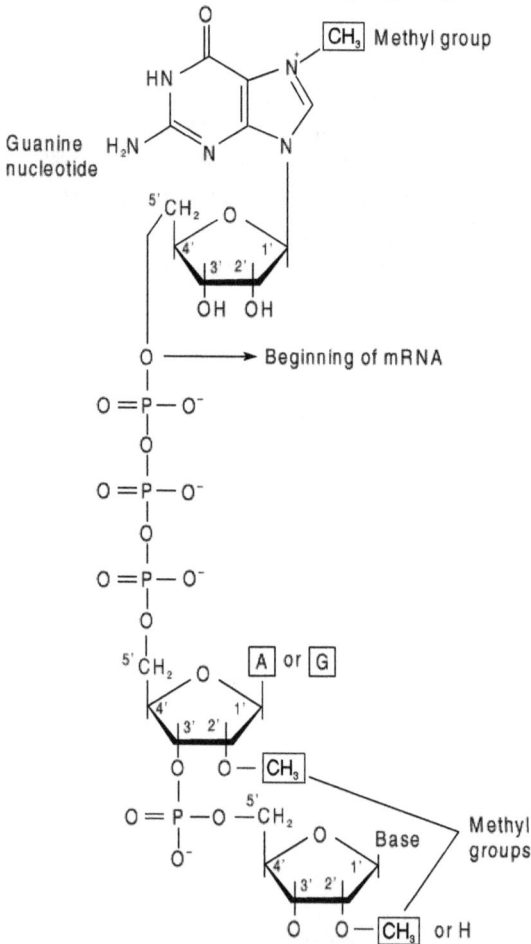

Figure 5.9   Molecular structure of mRNA molecule with 6-methyl guanosine cap

forming between the two end nucleotides, a triphosphate linkage forms. The triphosphate does not link the 3´-site of one nucleotide with the 5´-site of the next; instead, the two end nucleotides are linked through their 5´-sites. Thus, the capping of nucleotide "back in" to join the growing RNA chain. Strictly speaking, capping cannot be considered as a post-transcriptional process because it occurs before transcription is completed. It might be better to call it as "co-transcriptional process".

Capping has two important functions; first, it protects mRNA from degradation by RNases. Secondly caps are necessary for binding mRNA to the ribosome. Most RNA viruses cap their genomes and mRNAs except picornaviridae (it blocks 5´-end of its RNA by a protein), whereas orthomyxoviridae steals the preformed caps from host mRNA (cap snatching).

Polyadenylation

Most eukaryotic mRNAs have a chain of 40 to 200 adenine nucleotides attached to the 3´-end. This poly(A) tail is not transcribed from the DNA, but is rather added after transcription by the nuclear enzyme poly(A) polymerase. These tails may help stabilize mRNAs and facilitate their exit from the nucleus. After the mRNA enters the cytosol, the poly(A) tail is gradually shortened.

Most mRNAs contain poly(A). One famous exception is histone mRNA, which somehow manages to perform its function without a detectable poly(A) tail. One function of poly(A) is to protect the RNA from degradation by RNases.

A multi-subunit complex comprising a trimeric cleavage factor carries out cleavage and polyadenylation. These factors carry out the cleavage reaction. The enzyme polyadenylation polymerase (PAP) catalyses the addition of adenylate residue and several other uncharacterized components. Initial polyadenylation is slow because PAP dissociates after adding each adenylate residue. However, after a short oligo adenylate sequence has been generated, a further component, polyadenylate binding protein (PABP), attaches to the tail and increases the processivity of PAP. PABP, through an unknown mechanism, controls the maximum length of the polyadenylate tail.

Cleavage and Polyadenylation Complex

There are two enzymatic activities involved: an endonuclease and a poly(A) polymerase. However, there are several proteins known to be involved in this process. The proteins are shown in Table 5.1.

Table 5.1　Factors involved in polyadenylation

Abbreviation	Full Name	Function
CPSF	Cleavage and Polyadenylation Specificity factor	Required for both cleavage and polyadenylation, binds to AAUAAA
CF-I	Cleavage factor I	RNA binding protein, required for cleavage only
CF-II	Cleavage factor II	Required for cleavage only
CstF	Cleavage stimulation factor	Required for cleavage only, binds to GU/U region
PAP	Poly(A) polymerase	Catalyses poly(A) synthesis, also required for cleavage
PAB-II	Poly(A) binding protein II	Stimulates polyadenylation, required for lengthening of poly(A) tail

The exact function of only poly(A) polymerase is understood, while that other proteins is still unclear. In Figure 5.10, the poly(A) site is shown, with the signal AAUAAA and the downstream GU/U region. A complex of CPSF (the ball 1), CF I and II (the balls 2) and CstF (the ball 3) bind to these sequences.

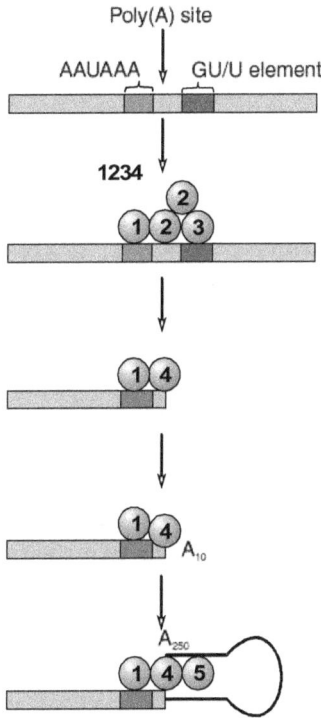

Figure 5.10   Molecular mechanism of capping

A cleavage occurs and CPSF remains and is joined by PAP (the ball 4). PAP begins the synthesis of poly(A), resulting in the addition of the first 10 As. Finally, PAB II joins the reaction, stimulating the synthesis of poly(A), extending the tail to about 200 A residues. In the text, two functions for poly(A) on mRNA are discussed, namely protection and translatability. In discussing these, the author points out rightly that the importance of these functions varies from system to system. In fact, since there are mRNAs that operate quite normally without poly(A) (such as the histone mRNAs) there is reason to doubt that these functions are universal. This is very different from the function of the 5´ cap which is now agreed to be required for the efficient translation of a eukaryotic mRNA.

Function of the Poly(A)

The stability of mRNA with or without poly(A) has been measured in a variety of systems. There is some evidence that poly(A) minus mRNA is less stable in certain systems. However, there is no general agreement about the meaning of these data.

Cap increases the translation of this message by some 300-fold, the presence of poly(A) increases the translatability of the message by about 20-fold. Although this is an effect, it is not a very great one. Added to this is the fact that some messages do not have poly(A) and you can see that the function is not at all clear.

RNA Splicing

RNA splicing involves removal of introns, the non-coding elements present in RNA molecule. Most genes of higher eukaryotes contain non-coding intervening sequences or introns separating the coding sequences or exons. The primary transcript contains the entire sequence of the gene and the non-coding sequences are spliced out during processing. Clearly, the splicing mechanism must be very precise, it must join exon sequences with accuracy to the single nucleotide to assure that codons in exons distal to introns are read in the correct reading frame. Accuracy to this degree would seem to require very precise splicing signals, presumably nucleotide sequences within introns and at the exon–intron junctions (Figure 5.11). As it turned out, there are three totally distinct types of intron excision from RNA transcripts.

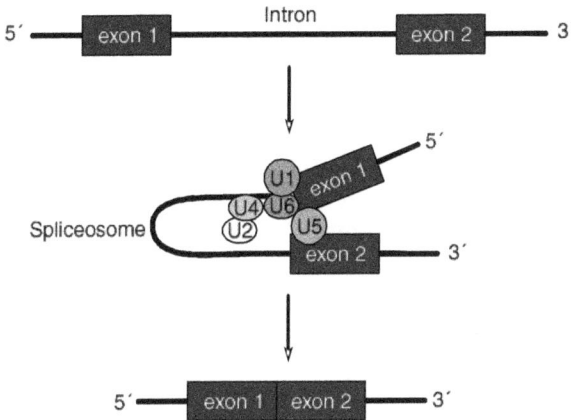

Figure 5.11   Molecules involved in intron splicing

Splicing tRNA precursors   The excision of introns from yeast tRNA precursors occurs in two stages. First, a nuclear membrane-bound splicing endonuclease makes two cuts precisely at the ends of the introns. Then in a fairly complex set of reactions, a splicing ligase joins the two halves of the tRNA to produce the natural form of the tRNA molecule. Introns are removed in multiple steps. Firstly, covalent modification of the substrate RNA is cleaved at a precise 5´-site followed by ligation. The first reaction is the addition of a phosphate group to the 5´-OH with the help of kinase enzyme using ATP as cofactor. Then, 5´-phosphate group is activated by the transfer of an AMP group to the terminus from an AMP-ligase intermediate. The 2´-3´ cyclic phosphate is opened by a cyclic phosphodiesterase activity that produces a 2´-phosphate and a free 3´-hydroxyl. The final ligation reactions occur via a nucleophilic attack of the free 3´-OH on the interior 5´-phosphate with the release of AMP. Finally, the 2´-phosphate group is removed by a phosphatase activity to yield the mature tRNA molecule. The overall two-stage mode of tRNA intron excision appears to occur in other organisms. In mammals, the reactions are not the same, splicing still occurs in two stages but the ligation reaction appears to directly join the 2′–3′ cyclic phosphate terminus to the 5′-OH terminus. The detailed procedure is not very clear.

Autocatalytic splicing of Tetrahymena   The autocatalytic excision of the intron in the *Tetrahymena* rRNA precursor requires no external energy source and no protein. Instead, it involves a series of phosphodiester bond transfers, with no bonds lost or gained in the process (Figure 5.12). The reaction requires a guanine nucleoside or nucleotide with a free 3′-OH group as a cofactor, plus a monovalent cation and a divalent cation. The requirement for the G-3-OH is compulsory and cannot be substituted by other factors or nucleosides.

The intron is excised by means of two phosphodiester bond transfers and the excised intron can subsequently circularize by means of another phosphodiester bond transfer. The autocatalytic circularization of the excised intron suggests that the self-splicing of these rRNA precursors reside primarily within the intron structure. A key point is that the autocatalytic splicing reactions are intramolecular in nature and are not concentration-dependent. Moreover, the RNA precursors are capable of forming an action centre in which the guanosine 3-OH cofactor binds.

Splicing with the help of spliceosome   In this mechanism, the introns are removed from the primary transcript in the nucleus, the exons are ligated to form the mRNA molecule with the help of special structures called spliceosomes. Spliceosome consists of the primary transcript, five small nuclear RNAs (U1, U2 U5 and U4/U6) and more than 50 proteins. Collectively these form a small nucleoprotein (snRNP) complex, sometimes called a snurp. Snurps are thought to position the RNA segments for the necessary splicing reaction.

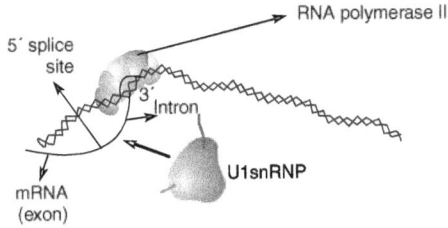

a. As soon as the 5′ splice site exits the transcription complex, a U1 snRNP binds to it.

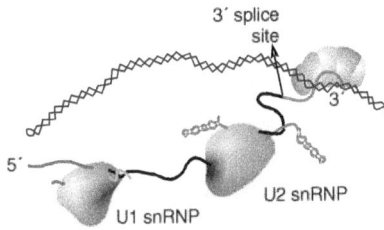

b. Next, a U2 snRNP binds to the branch site within the intron

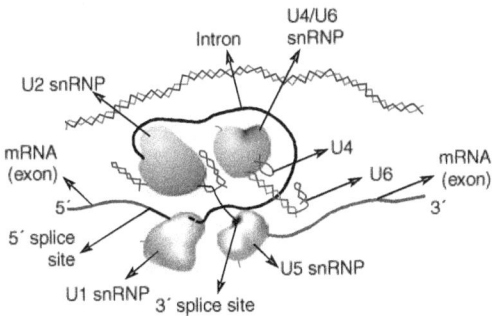

c. When the 3′ splice site emerges from the transcription complex, a U5 snRNP binds, and the complete spliceosome assembles around a U4/U6 snRNP

Figure 5.12 Complex group protein involved in rRNA splicing of the *Tetrahymena*

The splicing reaction starts with a cut at the junction of the 5′-exon and 3′-OH of intron by nucleophilic attack at the free 5′-terminal that forms a loop or lariat structure. A second cut is made at the junction of the intron with the 3′-exon. In this second reaction called transesterification reaction, the 3′-OH of the upstream exon attacks the 5′-phosphate at the downstream exon–introns boundary and the lariat structure containing the intron is released and hydrolysed. The 5′ and 3′ exons are ligated to form a continuous sequence.

The snRNAs and associated proteins are required for the formation of various structures and intermediates. U1 binds first by base pairing to the 5′-exon–intron boundary. U2 then binds by base pairing to the branch site and this exposes the nucleophilic A residue. A U5/U4/U6 complex next joins the spliceosome. An ATP-dependent protein-mediated unwinding results in the disruption of the base-paired U4/U6 complex with the release of U4. U6 is then able to interact first with U2 and then with U1. This process results in the formation of the loop or lariat structure. The U2/U6 complex cleaves the two ends and U6 is certainly essential for the spliceosome action. Interestingly, the introns are not removed in an order by manner but are randomly removed.

RNA EDITING

RNA editing is a process that results in changes in the RNA nucleotide sequence such that the RNA sequence differs from the DNA template from which it is transcribed. RNA editing also involves C to U change, addition of G or C residues, and conversion of U to A/G. RNA editing takes place with the help of guide RNAs (gRNAs). mRNAs are edited most frequently, whereas edits of tRNAs and rRNAs are extremely rare. The phenomenon of RNA editing was discovered in mid-1980 in the mitochondrial genes of *Trypanosoma* and *Leishmania* including the following.

1. Several genes having mutations that are supposed to render these genes inactive (loss of start stop/signal), are still active.

2. The sequences of mRNA molecules in several cases were such that they could not have been derived from the DNA sequence of the corresponding genes. Such genes are called as "cryptogenes".

The RNA editing mechanism is shown in Figure 5.13 and 5.14. It is observed that RNA editing mainly involves addition and deletion of U/C. RNA editing sometimes includes addition of up to 560 and deletion of up to 41 uridine residues. The addition or deletion is carried out with the help of a small RNA molecule of 40 bp length called guide RNA. It was also shown that the RNA apparently carry information for insertions and deletions at the tail region.

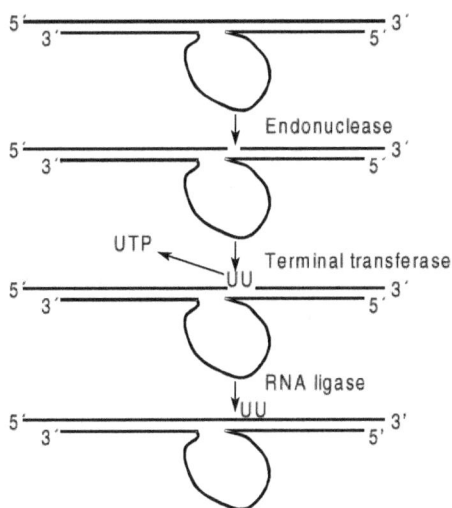

Figure 5.13 RNA editing mechanism

Like self-splicing of introns, RNA editing also involves transesterification, but with the help of guide RNA. First, the guide RNA aligns by base pairing itself with the unedited RNA, splits it into two and makes a new bond between one of the broken ends and the uridine at the tip of the tail of gRNA at its 3´-end. This reaction is facilitated by mitochondrial uridyl transferase activity. Next, the broken end of the other RNA segment, which is not involved, in addition to U forms a bond with this RNA segment. The tail of gRNA is now released for another round of transesterification.

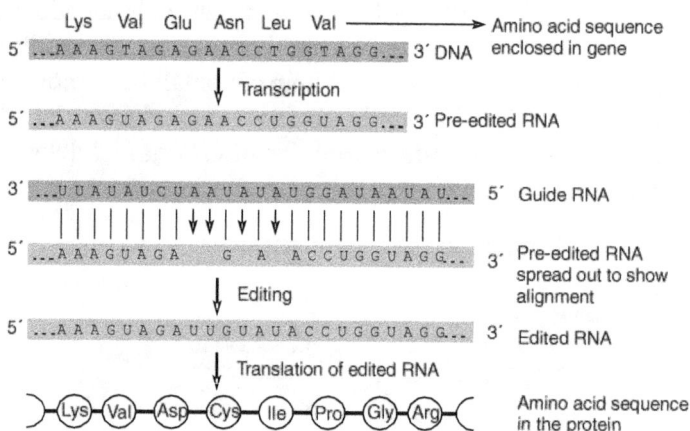

Figure 5.14   Simple diagram showing events and changes that occur during RNA editing

RNA editing challenges the role of genes as sole sites of genetic information. Hence, Thomas Cech called RNA editing as a molecular fossil because RNA editing is a living relic that provides a look backward at life's origin.

RNA editing is a post-transcriptional process that results in radical changes in the amino acid specified by a coding. Most of the editing events occur in the first two nucleotides of the coding sequences. RNA editing is observed in plant mitochondria, chloroplast and mammalian nucleus. The editing requires at least two steps, i.e., specificity and biochemical modification. The nucleotides located immediately upstream of the edited cytosine is most often a pyrimidine. RNA editing is divided into three types based upon the effect they have on the RNA.

1. *Simple editing*   It involves the conversion of single residue, e.g. C to U.

2. *Insertional editing*   It involves insertion of single nucleotides or small runs of nucleotides. This occurs by transcriptional strand slipping, e.g. G insertions.

3. *Pan-editing*   It involves insertion or deletion of multiple cytidine/uridine residues. This occurs with the help of external antisense guide RNA.

REVIEW QUESTIONS

1. What are the similarities and differences between DNA and RNA synthesis.

2. Describe the prokaryotic and eukaryotic RNA polymerases.

3. Describe the basic mechanism of transcription.

4. Write in detail about transcription factor.

5. Explain the process of precursor splicing in tRNA.

6. Describe the mechanism of autocatalytic splicing in *Tetrahymena.*

7. Explain the process of splicing with the help of splicesome.

8. Write in detail about RNA editing.

9. Write short notes on the following.

   i. Pribnow box

   ii. CAT box

iii.   TATA box

iv.   Enhancer

v.   Repressor

vi.   Upstream elements

vii.   Transcription termination

viii.   Eukaryotic promoters

ix.   Promoters for Pol III

x.   Promoters for Pol I

xi.   Transcription factors

xii.   mRNA capping

xiii.   Cleavage of introns

xiv.   Polyadenylation

xv.   Polyadenylation complex

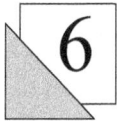

6

TRANSLATION

The process during which the genetic information is translated following the rules of genetic code into the sequence of amino acids in the polypeptide is called as translation or protein synthesis or polypeptide chain synthesis. Polypeptide chain synthesis requires a large number of components. These include

- ❑ Over 50 polypeptides in functional ribosomes
- ❑ At least 20 amino acid activating enzymes (aminoacyl-tRNA synthetases)
- ❑ 40 to 60 different tRNA molecules
- ❑ 9 soluble proteins
- ❑ Energy sources like GTP, ATP
- ❑ Amino acids

COMPONENTS OF TRANSLATION

Amino Acids

All the amino acids that eventually appear in the finished protein must be present at the time of protein synthesis. Even if one amino acid is not available, then the translation stops.

Transfer RNA (tRNA)

At least one specific type of tRNA is required per amino acid. In humans there are at least 50 species of tRNA, whereas bacteria contain 30 to 40 species. Because there are only 20 different amino acids commonly carried by tRNAs, some amino acids have more than one specific tRNA molecule. This is particularly true of those amino acids that are coded by several codons. The tRNA molecule has the following two important sites which are essential for protein synthesis.

❑ *Amino acid attachment site*   Each tRNA molecule has an attachment site for a specific amino acid at its 3′-end. When a tRNA has a covalently attached amino acid, it is said to be charged and the amino acid is called as uncharged.

❑ *Anticodon*   Each tRNA molecule also contains a 3-base nucleotide sequence, the anticodon, that recognizes a specific codon on the mRNA.

Aminoacyl-tRNA Synthetase

This family of enzymes is required for attachment of amino acids to their corresponding tRNAs. Each member of this family recognizes a specific amino acid and the tRNAs that correspond to that amino acid. Each aminoacyl-tRNA synthetase catalyses a two-step reaction that results in the covalent attachment of an amino acid to the 3′-end of its corresponding tRNA. The overall reaction requires ATP which is cleaved to AMP and PPi. The extreme specificity of the synthesis in recognizing these two structures is largely responsible for the high fidelity of translation.

Functionally Competent Ribosomes

Ribosomes are large complexes of protein and rRNA. They consist of two subunits. One large and one small whose relative sizes are generally given in Svedberg units. Prokaryotic and eukaryotic ribosomes are similar in structure and function but different in size.

Ribosomal RNA (rRNA)

Prokaryotic ribosomes contain 3 molecules of rRNA (5S, 23S and 16S RNA) and 53 protein whereas eukaryotic ribosomes contain 4 molecules of rRNA (5S, 28S, 5.8S and 18.5S RNA) and 80 proteins. The rRNAs have extensive regions of secondary structure arising from the base pairing of complementary sequences of nucleotides in different portions of the molecules. Ribosomal proteins are present in considerably greater number in eukaryotic ribosomes than in prokaryotes and play a number of roles in the structure and function of the ribosome and its interactions with other components of the translation system. The ribosome has two binding sites for tRNA molecules, the A and the P sites, each of which extends over both subunits. Together, they cover two neighbouring codons. During translation, the A site binds to an incoming aminoacyl-tRNA as directed by the codon currently occupying this site. This codon specifies the next amino acid to be added to the growing peptide chain. The P site codon is occupied by peptidyl-tRNA. This tRNA carries the chain of amino acids that have already been synthesized. In eukaryotic cells, the ribosomes are either "free" in the cytosol or in close association with ER. The RER-associated ribosomes are responsible for synthesizing proteins that are to be exported from the cell as well as those that are destined to become integrated into the plasma, ER, golgi membrane or lysosome. ATP and GTP are required as sources of energy. Cleavage of four high-energy bonds is required for the addition of one amino acid to the growing polypeptide chain. Two ATP are utilized in the aminoacyl-tRNA synthetase reaction (one for removing PPi and one for the subsequent hydrolysis of the PPi to inorganic phosphate by pyrophosphates) and two GTP are also utilized (one for binding the aminoacyl-tRNA to the A site, and one for the translation step).

GENETIC CODE

Genetic information is the sequence of bases in the DNA. Genetic information present in DNA is copied or expressed into RNA sequence. Genetic information that is expressed as an mRNA sequence must be translated from the four-letter language of nucleotides into the 21-letter language of proteins. The letters A, G, T and C correspond to the nucleotides found in DNA. They are organized into three-letter code words called codons.

A collection of these codons is called as genetic code. Genetic code is concerned with the processes involved in translating or decoding the information contained in the primary structure of DNA. The genetic code provides a foundation for explaining the way in which protein defects may cause genetic disease and for the diagnosis and perhaps treatment of these disorders. Some of the important properties of a genetic code are listed below.

Genetic code is degenerate   The occurrence of more than one codon per amino acid is called degeneracy. All amino acids except methionine and tryptophan have more than one codon, so that all the possible triplets have a meaning, despite there being 64 triplets and only 20 amino acids. Leucine, serine and arginine have six different codons. Isoleucine has three codons. The degeneracy in the genetic code is not at random: instead, it is highly ordered. Usually multiple codons specifying an amino acid differ by only one base, the third or 3′ base of codon. The degeneracy can be either partial or complete degeneracy. A complete degeneracy is a condition in which the third base position is of less significance. Any base present in this position will lead to same amino acid. But if the third position is occupied by one type of purine/pyrimidine, then they code for one type of amino acid and if the position is occupied by another type of purine/ pyrimidines, without any change in the first two positions, they code for another amino acid. Such a condition is called as partial degeneracy, e.g. CAU or CAC code for "His" whereas CAA and CAG code for "Gly".

Because of the degeneracy of the genetic code there must either be several different tRNAs that recognize the different codons specifying a given amino acid or the anticodon of a given tRNA must be able to base-pair with several different codons. Actually both of these occur. Several tRNAs exist for certain amino acids and some tRNAs recognize more than one codon. The hydrogen bonding between the bases in the anticodon of tRNA and the codon of mRNA appears to follow strict base-pairing rules only for the first two bases of the codon and is apparently less stringent, allowing what Crick has called wobble at this site.

Codon is triplet   Twenty different amino acids are incorporated during translation. Thus, at least 20 different codons must be formed using the four symbols (bases) available in the mRNA. Two bases per codon would yield only 4^2 or 16 possible codons which is clearly not enough. Three bases per codon yield 4^3 or 64 possible codons–

apparently excess. It is now known that each codon consists of a sequence of three nucleotides, i.e., it is a triplet code. The deciphering of the genetic code depended heavily on the chemical synthesis of nucleotide polymer, particularly triplets in repeated sequence.

Code contains punctuation codons   Three codons do not code for specific amino acids. These codons are called as nonsense codons. These nonsense codons cause termination of protein synthesis. Thus they are called as termination codons. Similarly, AUG codon codes for starting of the gene and position where translation should begin. AUG codes for the initiation codon and because it codes for methionine, almost all newly synthesized polypeptides have this amino acid at the start.

Codon is commaless   There are no commas or some specific nucleotide sequences to separate the codons, i.e., CCCAAAUUUGGG has four code words and upon translation we have a tetrapeptide chain of pro-lys-phen-gly. So all the letters are used to code for one or other amino acid.

Wobble Hypothesis

The genetic code is a degenerate code, meaning that a given amino acid may have more than one codon. Therefore for these eight amino acids, the codon need only be read in the first two positions because the same amino acid will be represented regardless of the third base of the codon. These eight groups of codons are termed unmixed families of codons. An unmixed family is the four codons beginning with the same two bases that specify single amino acids. For example, the codon family GUX codes for valine.

Table 6.1 Genetic code table

5′ Base	Middle base				3′ Base
	U	C	A	G	
U	UUU ⎤ Phe UUC ⎦ UUA ⎤ Leu UUG ⎦	UCU ⎤ UCC ⎬ Ser UCA ⎪ UCG ⎦	UAU ⎤ Tyr UAC ⎦ UAA ochre UAG amber	UGU ⎤ Cys UGC ⎦ UGA opal UGG Try	U ⎤ Pyrimidines C ⎦ A ⎤ Purines G ⎦
C	CUU ⎤ CUC ⎬ Leu CUA ⎪ CUG ⎦	CCU ⎤ CCC ⎬ Pro CCA ⎪ CCG ⎦	CAU ⎤ His CAC ⎦ CAA ⎤ Gln CAG ⎦	CGU ⎤ CGC ⎬ Arg CGA ⎪ CGG ⎦	U C A G
A	AUU ⎤ Ile AUC ⎬ AUA ⎦ AUG Met	ACU ⎤ ACC ⎬ Thr ACA ⎪ ACG ⎦	AAU ⎤ Asn AAC ⎦ AAA ⎤ Lys AAG ⎦	AGU ⎤ Ser AGC ⎦ AGA ⎤ Arg AGG ⎦	U C A G
G	GUU ⎤ GUC ⎬ Val GUA ⎪ GUG ⎦	GCU ⎤ GCC ⎬ Ala GCA ⎪ GCG ⎦	GAU ⎤ Asn GAC ⎦ GAA ⎤ Glu GAG ⎦	GGU ⎤ GGC ⎬ Gly GGA ⎪ GGG ⎦	U C A G

Universality of the Code

A vast amount of data suggest that the genetic code is universal. It was postulated that the genetic code must be frozen and unable to evolve because a change in a codon meaning would cause almost every protein in the cell to be altered. Experiments with cell-free system with mammalian extract yielded same results as that of *E. coli*. The major exception to the universality of the code occurs in mitochondria of humans, yeast and several other species where UGA is a tryptophan codon. UGA is a termination codon in the non-mitochondrial system. Also, in yeast mitochondria, CUA specifies threonine instead of the usual leucine and in mammalian mitochondria, AUA specifies methionine instead of the usual isoleucine. Still more surprising is the discovery that the meaning of a codon can vary from gene to gene in the same organism. In human nuclear genes UGA is frequently used as a termination codon in accordance with its standard meaning. However, in at least two human genes, for the enzymes glutathione peroxidase and iodothyronine 5´-deiodinase, UGA specifies the unusual amino acid called selenocysteine (a cysteine in which the sulphur atom is replaced by selenium). It appears that the mRNA transcribed from these genes have special stem-loop structures in their trailer regions, and that these stem-loops play some role during translation, ensuring that the UGA triplet is recognized as a selenocysteine codon rather than a termination signal.

EUKARYOTIC TRANSLATION

Translation, like transcription, is divided into three phases or steps. The mechanism of eukaryotic translation and the molecules involved are shown in Figure 6.1.

Initiation

Initiation of protein synthesis requires an mRNA molecule to be selected for translation by a ribosome. Once the mRNA binds to the ribosome, translation begins. This process involves tRNA, rRNA, mRNA and at least 10 eukaryotic initiation factors (eIFs). Initiation consists of four steps.

Ribosomal dissociation   Two initiation factors eIF3 and eIF1A, bind to the newly dissociated 40S ribosomal subunit. This delays its re-association with 60S subunit and allows other translational initiation factors to associate with the 40S subunits.

Formation of the 43S pre-initiation complex   The first step in this process involves the binding of GTP by eIF2. This binary complex then binds to met-tRNA, a tRNA specifically involved in binding to the initiation codon AUF. This tertiary complex binds to the 40S ribosomal subunit to form the 43S pre-initiation complex which is stabilized by association with eIF3 and eIF1A.

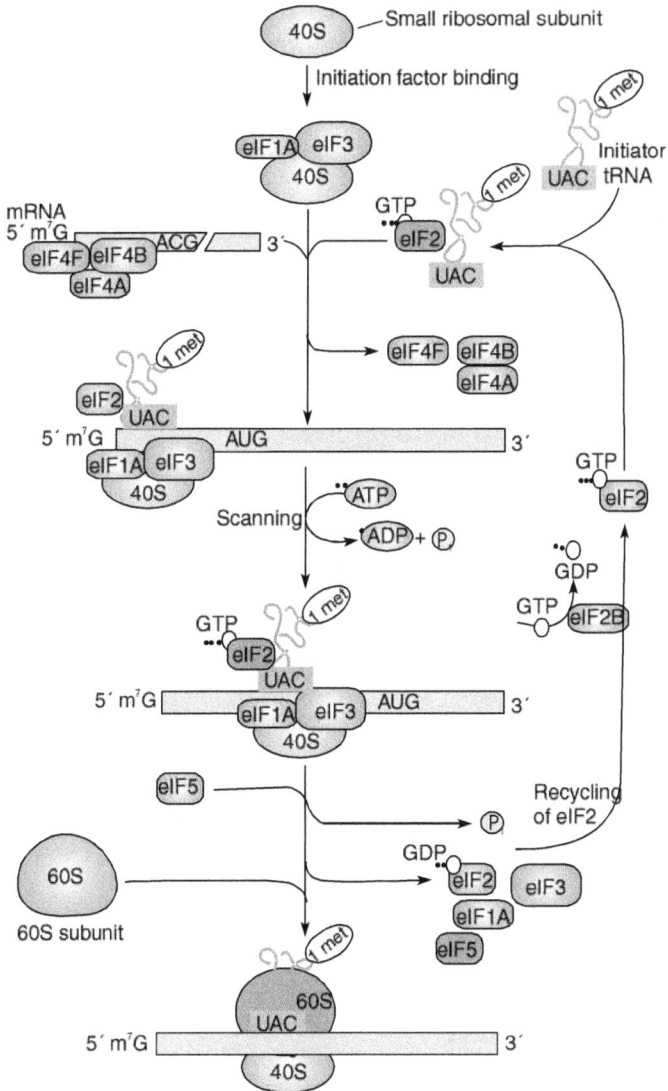

Figure 6.1 Translation mechanism with molecules involved in eukaryotic translation

Formation of the 43S initiation complex The 5′ terminals of most mRNA molecules in eukaryotic cells are "capped". This methyl-guanosyl triphosphate cap facilitates the binding of mRNA to the 43S pre-initiation complex. A cap-binding protein complex, (eIF4F) binds to the cap through the 4E protein. The association of mRNA with the 43S pre-initiation complex to form the 48S initiation complex requires ATP hydrolysis. Following the association of the 43S pre-initiation complex with the mRNA cap and reduction of the secondary structure near the 5′ end of the mRNA, the complex scans the mRNA for suitable initiation codon. Generally this is the 5′-most AUG. But the precise initiation codon is determined by the so-called "kozak" consensus sequences that surround the AUG in eukaryotes.

Formation of the 80S initiation complex The binding of the 60S ribosomal subunit to the 48S initiation complex involves the hydrolysis of the GTP bound to eIF2 by eIF5. This reaction results in the release of the initiation factors bound to the 48S initiation complex and the rapid association of the 40S and 60S subunits to form the 80S ribosome.

Elongation

Elongation is a cyclic process involving several steps (Figure 6.2) catalysed by proteins called elongation factors eEF. These steps are

1. Binding of aminoacyl-tRNA to the A site
2. Peptide bond formation
3. Translocation

Binding of aminoacyl-tRNA to the A site In the complete 80S ribosome formed, during the process of initiation, the A site (aminoacyl or acceptor site) is free. The binding of the proper aminoacyl-tRNA in the A site requires

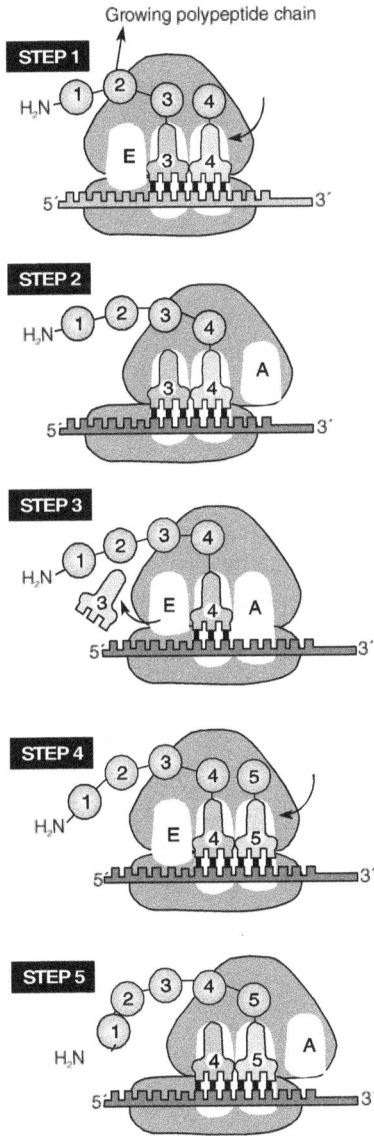

Figure 6.2 Steps that occur during elongation

proper codon recognition—elongation factor eEF-1x-GDP and phosphate.

Peptide bond formation　The x-amino group of the new aminoacyl-tRNA in the A site carries out a nucleophilic attack on the esterified carboxyl group of the peptidyl tRNA occupying the P site. This reaction is catalysed by a peptidyl transferase, a component of the 28S RNA of the 60S ribosomal subunit.

Translocation　Upon removal of the peptidyl moiety from the tRNA in the P site, elongation factor (eEF2) and GTP are responsible for the translocation of the newly formed peptidyl tRNA at the A site into the empty P site. The GTP required for eEF2 is hydrolysed to GDP and phosphate during the translocation process. The translocation of the newly formed peptidyl tRNA and its corresponding codon into the P site frees the A site for another cycle of aminoacyl-tRNA codon recognition and elongation. The energy required for the formation of one peptide bond includes the hydrolysis of 2 ATP and 2 GTP molecules. This process occurs rapidly. A eukaryotic ribosome can incorporate as many as six amino acids per second, whereas 18 amino acids are incorporated per second by prokaryotic ribosomes. Thus, the process of peptide synthesis occurs with great speed and accuracy until a termination codon is reached in prokaryotes.

Termination

In comparison to initiation and elongation, termination is relatively a simple process. Multiple cycles of elongation occur culminating in polymerization of the specific amino acids into a protein molecule. There is no tRNA with an anticodon capable of recognizing such a termination signal. Releasing factors (eRF) are capable of recognizing termination signal residues in the A site. The releasing factor, in conjugation with GTP and the peptidyl transferases, promotes the hydrolysis of the bond between the peptide and the tRNA occupying the P site. The ribosome dissociates into 40S and 60S subunits.

PROKARYOTIC TRANSLATION

The pathway of protein synthesis is called translation because the "language" of the nucleotide sequence on the mRNA is translated into the language of an amino acid sequence. The mRNA is translated from the 5′ end to its 3′ end producing a protein.

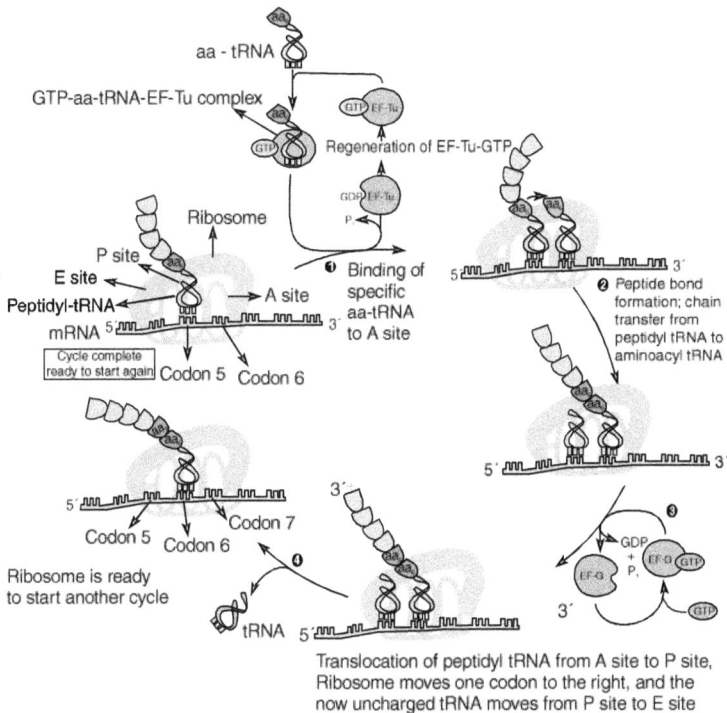

Figure 6.3 Sequence of events that occur during translation in prokaryotes

Prokaryotic mRNAs often have several coding regions, that is, they are polycistronic. Each coding region has its own initiation codon and produces a separate species of polypeptide. Prokaryotic translation resembles that of eukaryotic translation in most of the detail except with minor changes (Figure 6.3).

Initiation

In prokaryotes, the ribosome 70S instead of 80S is used in translation machinery. In prokaryotes, the first amino acid, the one that initiates protein synthesis, is always the same. N-formyl methionine (Met). N-formyl methionyl-tRNA forms in two steps. First, ordinary methionine binds to a special tRNA, then the methionine on the tRNA receives the formyl group. This unique amino acid participates only in initiation. It never goes into the interior of a polypeptide. In *E. coli,* a sequence of nucleotide bases (5′-UAGGAGG -3′) known as the Shine–Dalgarno sequence is located 6 to 10 bases upstream of the AUG codon on the mRNA molecule, that is, near its 5′-end. The 16S ribosomal RNA component of the 30S ribosomal subunit has a nucleotide sequence near its 3′-end that is complementary to all part of the Shine–Dalgarno sequence. Therefore, the mRNA 5′-end and the 3′-end of the 16S ribosomal RNA can form complementary base pairs, thus facilitating the binding and positioning of the mRNA on the 30S ribosomal subunit.

Elongation

During elongation, prokaryotes use EF-TU instead of eEF-1α for catalysis, whereas for translocation they use EF-G instead of eEF-2.

Interestingly, many ribosomes can translate the same mRNA molecule simultaneously. Because of their relatively large size, the ribosome particles cannot attach to an mRNA any closer than 80 nucleotides apart. Multiple ribosomes on the same mRNA molecule form a polyribosome or polysome.

POST-TRANSLATIONAL MODIFICATIONS

Many polypeptide chains are covalently modified, either while they are still attached to the ribosome or after their synthesis has been completed. Because the modification occurs after translation is initiated, they are called post-translational modifications. These modifications may include removal of part of the translated sequence or the covalent addition of the translated sequence or the covalent addition of one or more chemical groups required for protein activity.

Trimming Many proteins destined for secretion from the cell are initially made as large, precursor molecules that are not functionally active. Portions of the protein chain must be removed by specialized endoproteases, resulting in the release of an active molecule. The cellular site of the cleavage reaction depends on the protein to be modified. For example, some precursor proteins are cleaved in the ER or the golgi complex. Zymogens are inactive precursors of secreted enzymes. They become activated through cleavage once they have reached their proper site of action.

Covalent alterations Proteins, both enzymatic and structural, may be activated or inactivated by covalent attachment of a variety of chemical groups.

Phosphorylation   Phosphorylation occurs on the hydroxyl groups of serine, threonine or less frequently on the tyrosine residues in a protein. This phosphorylation is catalysed by one of a family of protein kinase enzymes, and phosphorylation may increase or decrease the functional activity of the protein.

Glycosylation   Many of the proteins that are destined to become part of a plasma membrane or secreted from the cell have carbohydrate chain attached to serine or threonine hydroxyl groups (O-linked) or asparagine (N-linked). The stepwise addition of sugar occurs in the ER and the golgi complex.

Hydroxylation   Proline and lysine residues of the X-chain of collagen are extensively hydroxylated in the ER. In this process an OH group is added to the amino acid. This hydroxylation reaction requires molecular oxygen and a reducing agent such as vitamin C.

Sulphation   This involves addition of sulphur group to tyrosine residue.

Myristylation   This involves addition of acetyl group to the N-terminal of protein or asparagine.

REVIEW QUESTIONS

1. Describe the differences and similarities between transcription and translation.

2. Describe the mechanism of translation.

3. Describe the structure of tRNA.

4. Explain in detail the process of eukaryotic translation.

5. Explain in detail the process of prokaryotic translation.

6. Describe in detail post-translational modification.

7

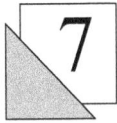

REGULATION OF GENE EXPRESSION

GENE EXPRESSION IN PROKARYOTES

Bacteria such as *E. coli* are exposed to a wide variety of environmental conditions. Natural selection will preserve those organisms that have evolved ways of adapting to the wide range of environmental conditions encountered during their evolution. The adaptability of bacteria and other prokaryotes depends on their ability to turn on and off the expression of specific sets of genes in response to the specific demands of the environment. The expression of a particular gene is turned on when the products of these genes are needed for growth in a given environment. Their expression is turned off when their products are no longer needed for growth in the existing environment.

The synthesis of gene transcripts and translation products requires the expenditure of considerable energy. By "turning off" the expression of genes when their products are not needed, an organism can avoid wasting energy and can utilize the conserved energy to synthesize products that maximize the growth rate. For example, if a given metabolite is not present, enzymes for its breakdown are not necessary, and synthesizing these enzymes is wasteful. If a cell produces enzymes for the degradation of a particular carbon source only when this carbon source is present in the environment, the enzyme system is known as an inducible system. In contrast, the enzymes in many synthetic pathways are in low concentration or absent when an adequate quantity of the end product of the pathway is already available to the cell. That is, if the cell encounters an abundance of the amino acids, e.g. tryptophan in the environment or if it is over-produced, the cell stops the manufacture of tryptophan until a need arises again. A repressible system is a system of enzymes whose presence is repressed, stopping the production of the end product when it is no longer needed. Repressible systems are repressed by an excess of the end product of their synthetic pathway. Bacteria regulate or alter the gene expression either by using positive or negative regulation. Positive and negative regulation look similar in most aspects, but the fundamental difference is what type of molecule, i.e., repressor alone, e.g. *lac* operon or repressor along with inducer/end product, e.g. *trp* operon, is binding to promoter, and also whether the molecule is increasing (inducer) or decreasing (repressor) the gene expression.

Inducer   The substance which induces gene expression or protein synthesis is called as inducer. The phenomenon is called as induction.

Repressor   The substance which stops or represses the expression of specific genes is called as repressor and the phenomenon is called as repression.

If the presence of a specific regulatory element enhances the level of gene expression, then that is called as positive regulation and the molecule is called as activator, e.g. *lac* operon.

If the presence of a specific regulatory molecule will reduce the gene expression, then it is called as negative regulation and the molecule is called as repressor.

Jacob and Monad in 1961 proposed a model which explains how a bacteria metabolizes lactose and tryptone, and the model is called as Jacob–Monad model. Bacteria use a positive control to metabolize lactose and negative control or regulation to metabolize tryptone.

OPERON

An operon is a set of genes which are linked and are under the control of one promoter or operator. These genes accomplish one single task. An operon basically consists of two categories of genes.

1. *Structural genes*   These genes are segments of DNA, which code for functional peptides, or enzymes or proteins. Proteins of structural genes directly interact with the inducer or accomplish a single task.

2. *Control genes*   These are genes which are primarily responsible for controlling the structural genes by producing an inducer or repressor substance. There are basically three types of genes.

i. *Regulator genes*   The regulator gene (r) produces some specific enzymes which act as repressor substances. This repressor binds to the operator gene and thus stops the expression of structural genes.

ii. *Promoter gene*   The promoter gene (p) is the DNA segment at which RNA polymerase binds. It initiates the transcription of the structural genes.

iii. *Operator gene*   The operator gene (o) is the segment of DNA which exercises a control over transcription. It lies close to the structural gene and the repressor binds to it.

Function of Operon

In the absence of an inducer (lactose) the regulator gene produces a protein repressor which binds strongly to the operator site and prevents its transcription. As a result, the structural genes do not produce an mRNA. But in the presence of inducer, the inducer binds to repressor. Thus the operator is free and induces the synthesis of RNA from structural genes. Induction should not be confused with enzyme activation in which the binding of a small molecule to an enzyme increases the activity of the enzyme.

❑ The complete contiguous unit, including the structural gene or genes, the operator and the promoter is called an operon.

❑ The only essential difference between inducible operons and repressible operons lies in whether the naked repressor or the repressor–effector molecule complex is active in binding to the operator. In the case of an inducible operon, the free repressor binds to the operator, turning off transcription. When the effector molecule is present, it binds to the repressor, releasing the repressor from the operator, that is, the repressor–inducer complex cannot bind to the operator. Thus, the addition of inducer turns on or induces the transcription of the structural genes in the operon. In the case of a repressible operon, the situation is just reversed. The free repressor cannot bind to the operator, only the repressor–effector molecule complex is active in binding to the operator.

Thus, transcription of the structural genes in a repressible operon is turned on in the absence of and turned off in the presence of the effector molecule. Except for this difference in the operator-binding behaviour of the repressor, inducible and repressible operons are comparable.

Lac OPERON

The lac operon model (Figure 7.1) was proposed by Jacob and Monad in 1961 to suggest the possible way of answering how a bacteria metabolizes the lactose. *Lac* operon functions only when lactose is present in the media (Figure 7.2). *Lac* operon basically consists of the following.

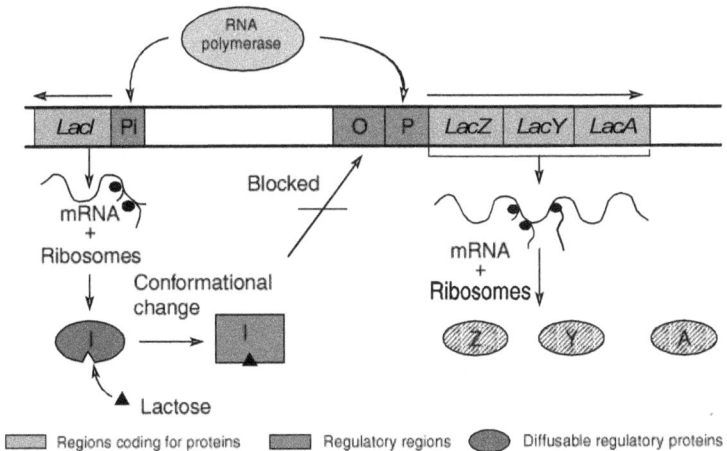

Figure 7.1 *lac* operon gene structure and its function

1. *Regulator gene* The regulator gene codes for a protein (360 AA), which is called as repressor. If the lactose is present in the media, lactose binds to the repressor. The repressor fails to bind the operator gene. If lactose is absent, the repressor binds to the operator gene. *Lac* repressor is a tetramer consisting of identical subunits.

2. *Promoter gene*   Promoter gene is present between the regulator and operator gene. RNA polymerase binds to promoter gene and initiates the transcription of the structural genes. RNA polymerase can initiate the transcription if the operator gene is free, i.e, repressor is not bound to the operator. A binding site of another protein called catabolite activator protein (CAP), functions such that the *lac* operon is not transcribed in the presence of glucose at concentrations sufficient to support optimal growth.

3. *Operator gene*   The operator gene is the segment of DNA to which a repressor binds. The operator gene is present in between the promoter and structural gene. It is 28 bp in length and is a palindrome.

4. *Structural genes*   There are three structural genes in the *lac* operon. These genes are named as cistron-Z, cistron-Y and cistron-A. Cistron-Z codes for an enzyme called β-galactosidase. This enzyme cleaves the lactose into galactose and glucose. Cistron-Y codes for an enzyme called galactoside permease. This protein is located on the cell wall and facilitates the transfer of lactose from outside into the cell. Cistron-A codes for galactoside transacetylase.

Function of *lac* Operon

In the presence of lactose, the lactose binds to the repressor and prevents it from binding to the promoter by inducing structural changes in the repressor. Then the RNA polymerase binds to the promoter and starts the transcription of the structural gene.

The structural gene is expressed as a single transcript. But different ribosomes bind to RNA transcript each at the starting of each cistron. Thus it makes the protein. These proteins then break down the lactose into glucose units. This gene expression continues as long as lactose is present in the media. Once the lactose gets depleted in the media, the repressor binds to the operator gene and stops the transcription of the structural gene.

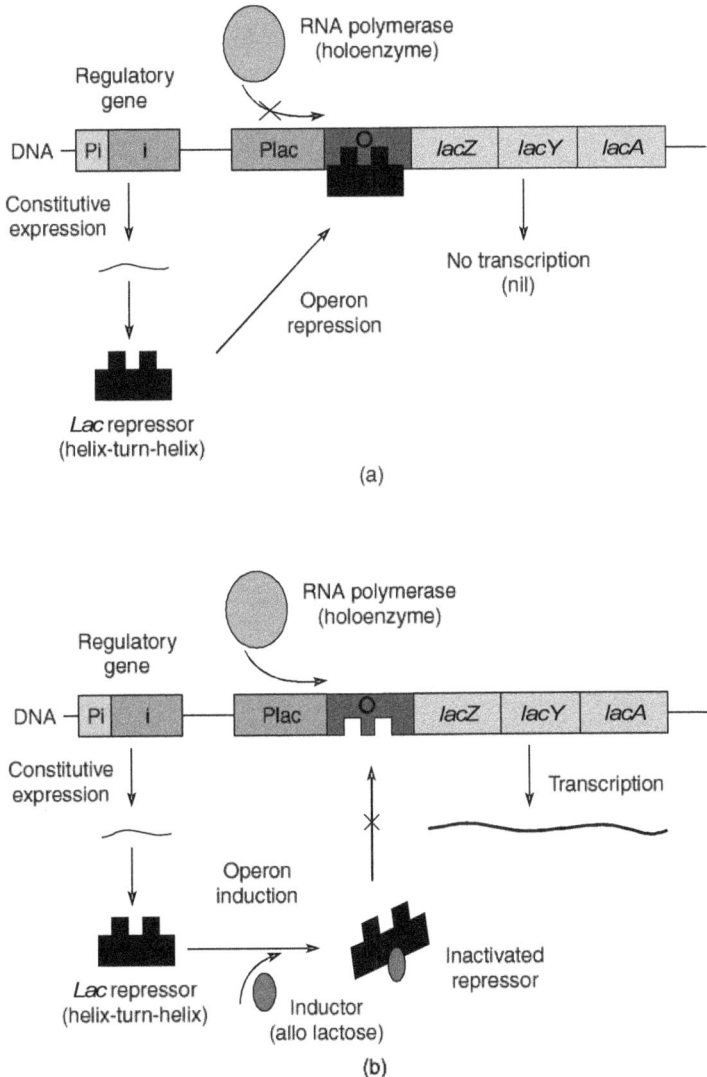

Figure 7.2   *Lac* operon (a) in the absence and (b) in the presence of lactose

The basic reason for this on and off of the gene is to ensure that the bacteria makes only the protein responsible/required at the particular time, thus avoiding the wastage of resources and energy on preparing proteins that are not required at that moment. As long as the lactose is absent in the media the repressor binds to the operator gene. Thus the transcription of the structural gene does not take place.

Positive Control of *lac* Operon

Interestingly, if bacteria encounter two sugars, e.g. glucose and lactose, then *E. coli* cells will keep the *lac* operon turned down as long as glucose is present. It looks appropriate if glucose is still available, because, *E. coli* cells metabolize glucose more easily than lactose, it would therefore be wasteful for them to activate the *lac* operon in the presence of glucose. The selection of glucose metabolism against the use of other energy sources has long been attributed to the influence of some breakdown product or catabolite of glucose. It is therefore known as catabolite repression. The ideal positive controller of the *lac* operon would be a substance that senses the lack of glucose and responds by activating the lac promoter so that RNA polymerase could bind and transcribe the structural genes. One substance that responds to glucose concentration is a nucleotide called cyclic AMP (cAMP). As the level of glucose drops, the concentration of cAMP rises. Then CAP (catabolite activator protein) binds to cAMP and the resulting complex binds to the *lac* control region and helps RNA polymerase to bind there. X-ray diffraction studies of the CAP/cAMP/promoter complex has shown that when CAP/cAMP binds, it causes the DNA to bend by 90°. This bending apparently helps RNA polymerase to separate the two DNA strands, forming an open promoter complex. CAP/cAMP may also stimulate transcription by direct protein–protein contact between CAP and RNA polymerase.

Mutations in *lac* operon lead to the formation of two types of mutants.

1. *Cryptic mutants* These mutants cannot metabolize the lactose, as they lack the ability to produce the enzyme galactoside permease.

2. *Constitutive mutants* Constitutive mutants are mutants in which the three *lac* operon genes are transcribed at all times, i.e., they are not turned off even in the absence of lactose. These mutants arise by mutations in the repressor gene due to which the system turns neither on nor off.

trp OPERON

The *trp* operon (Figure 7.3) consists of the genes for the enzymes that are needed to make the amino acid tryptophan. The *trp*

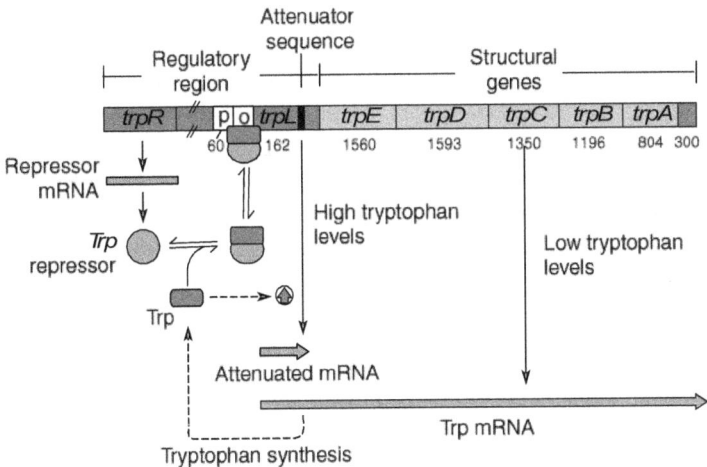

Figure 7.3 *trp* operon gene structure and function

operon on the other hand codes for anabolic enzymes that build up a substance. Such operons are generally turned off by that substance. When the tryptophan concentration is high, the product of the *trp* operon is not needed any longer and so the *trp* operon also exhibits an extra level of control not seen in the *lac* operon.

1. *Repressor*   The *trp* repressor binds to the operator site. The core-binding site is a palindrome of 18 bp. The *trp* repressor binds to tryptophan if present in the medium. Repressor can only bind to the operator when it is complexed with tryptophan. The repressor is a dimer of two subunits, which have structural affinity towards the operator site.

2. *Structural gene*   The *trp* operon structural gene consists of 5 genes (*trpE, trpP, trpC, trpB* and *trpA*). The operon encodes a single transcription unit which is 7 kb in length.

Function of *trp* Operon

The first two genes *trpE* and *trpD* code for polypeptides that make up the first enzyme in the pathway. The third gene, *trpC* codes for a single polypeptide enzyme that catalyses the second step. The last genes *trpB* and *trpA* code for the two polypeptides of the enzyme that carry out the third and last step in the pathway. When tryptophan is present in the media, it binds to the repressor and induces conformational change. Thus the repressor binds to the operator gene. This binding prevents the RNA polymerase from transcribing the structural genes (Figure 7.4). When tryptophan is absent in the media, the repressor is free and does not undergo any conformational changes. Thus the repressor does not bind to the operator gene. This facilitates the binding of RNA polymerase and transcription of structural gene.

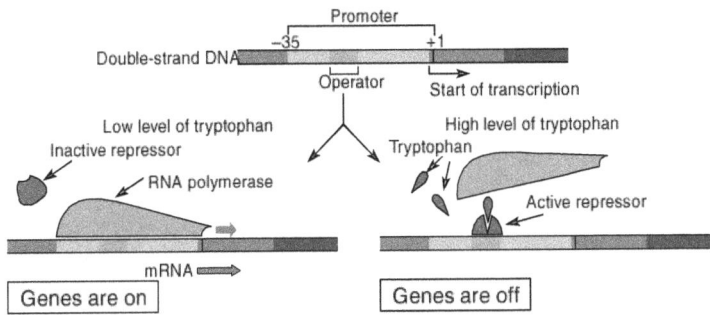

Figure 7.4   Model showing how tryptophan interacts with *trp* operon and inactivates the RNA polymerase binding

Control of the *trp* Operon by Attenuation

In addition to the standard, negative control scheme (*lac* operon) the *trp* operon employs another mechanism of control called "attenuation". Repression of the *trp* operon is considerably weak. Thus considerable transcription of the *trp* operon can occur even in the presence of repressor. In fact, in attenuator mutants where only repression can operate, the fully repressed level of transcription is 70-fold lower than the fully depressed level. The attenuation system permits another 10-fold control over the operon's activity. This is valuable because synthesis of tryptophan requires considerable energy.

In the *trp* operon, an attenuator region lies between the operator and the first structural gene. The messenger RNA transcribed from the attenuator region termed the leader transcribe has four (1, 2, 3 and 4) sub-regions of the messenger RNA and has base sequences that are complementary to each other so that three different stem-loop structures can form in the messenger RNA. Depending on circumstances, regions 1–2 and 3–4 can form two stem-loop structures or region 2–3 can form a single stem-loop. When one stem-loop structure is formed, the others are preempted.

When excess of tryptophan is present in the medium, the transcription of the gene continues past the leader peptide. Three different outcomes can take place. The moving ribosome overlaps regions 1 and 2 of the transcript and allows the stem-loop for 3–4 to form. This stem-loop structure causes transcription to terminate. Hence this stem-loop (3–4) structure is called as the terminator or attenuator stem. When there is no tryptophan, the *trp* operon must be activated. Accordingly, there must be a means of overriding attenuation. In this case, the stem-loop form will also form, but instead of forming two stem-loops (1–2 and 3–4) only a single stem-loop (2–3) is formed. In this configuration, transcription is not terminated so that eventually, the whole operon is transcribed and translated raising the level of tryptophan in the cell. The stem-loop 2–3 structure is referred to as the preemptor stem.

GENE EXPRESSION IN EUKARYOTES

Genes in prokaryotes are apparently continuously "turned on" and must be repressed if activity is to be controlled. In general genes of eukaryotes are apparently "turned off" and must be activated. At any time, only 2–15% of total genes are expressed. It means some sort of regulation is occurring.

The operon system of gene regulation has been proved to be inapplicable to eukaryotes. In higher eukaryotes, it does seem very clear that operons are not important, if at all they exist. Although there is evidence for operons or operon-like units in lower eukaryotes, they are polygenic and are cleaved to produce monogenic mRNAs. In eukaryotes, metabolically related genes may be scattered throughout the genome. However, groups of eukaryotic genes involved in the same pathway or function can be induced simultaneously by having common enhancers that respond to the same specific transcription factors. Such a group of genes is called a synexpression group.

Moreover, in eukaryotic cells, compartmentalization is present. The nucleus contains most of the genetic material. The compartmentalization has a consequence for gene expression. The expression of genes is regulated and the coordinated regulation of sequential pathways of gene expression is primarily responsible for the diversity of cell phenotypes that unfold during the development of a higher plant or animal.

Regulatory processes in eukaryotes act, or might act, to control gene expression at various levels. Various potential control points recognized in eukaryotes are regulation at the level of gene structure, initiation of transcription, processing transcript and transport to cytoplasm, and translation of mRNA.

REGULATION AT THE LEVEL OF GENE STRUCTURE

Chromatin Remodelling

For transcription to take place in eukaryotes, the DNA must be available for the pre-initiation complex to form, with its RNA polymerase and general transcription factors. It appears that DNA wrapped around nucleosomes is often not accessible for the formation of the pre-initiation complex, but is available for recognition by transcription-activating proteins, (specific transcription factors). One model of initiation of transcription by genes whose promoters are wrapped around nucleosomes is for specific transcription factors to recruit chromatin-remodelling protein, e.g. histones, acetyl transferase, ATP-dependent chromatin-remodelling proteins. Thus, the presence of one or more specific transcription by recruiting chromatin-remodelling proteins allows the RNA polymerase access to the promoter (Figure 7.5).

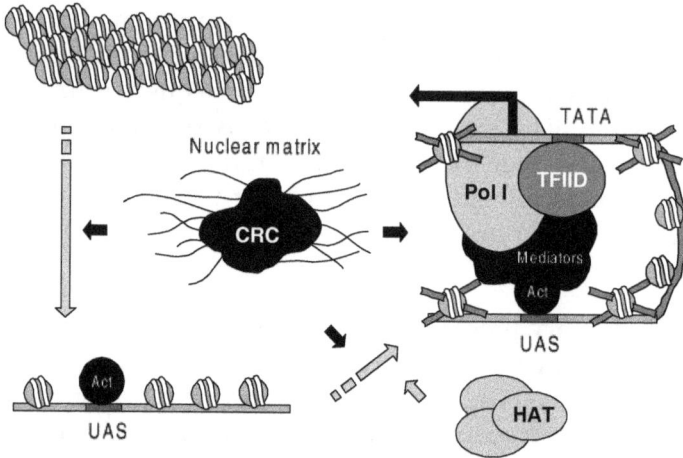

Figure 7.5   Chromatin remodelling and transcription

Twisting

In the case of twisting, the regulatory proteins are thought to bind to some altered form of DNA (e.g. left-handed DNA or single-stranded DNA) or regulatory protein that has an enzymatic activity that alters DNA conformation. Gene activation would be a consequence of conformational change propagated through the DNA, which allows other proteins to bind and begin transcription. All known or suspected regulatory proteins isolated from prokaryotes and eukaryotes are believed to recognize ordinary helical DNA.

Sliding or Tracking

Sliding is also referred to as tracking. This mechanism imagines that a protein recognizes a specific site on DNA and then moves along the DNA to another specific sequence where perhaps by interacting with another protein, it initiates transcription. Regulatory proteins are not known to act by sliding but this process is important for many other precise DNA transactions. Type I restriction enzymes bind their unique recognition sequences and cleave the DNA at sites that can be thousands of base pairs away. This is possible only through tracking.

Oozing

In this case, binding of a regulatory protein to its operator helps binding of another protein to adjacent sequences which in turn helps another to bind next to it, until a procession of proteins has oozed out from the control sequences to the gene where transcription is initiated. Oozing effect can function only for a few hundred base pairs.

CIS-ACTING ELEMENTS

Eukaryotic genes are regulated by promoter elements located just upstream (5′) from the transcription–initiation sites in a manner quite similar to the regulation of prokaryotic genes. In addition to the nearby promoters, many eukaryotic genes are also regulated by more distant *cis*-acting elements called enhancers and silencers.

Enhancers increase the transcription and are independent of orientation. Enhancers can act over relatively large distances. They are present several thousand bp from the regulated genes and are relatively large elements up to several hundred nucleotide pairs in length. They sometimes contain repeated sequences that have partial enhancer activity by themselves. Most enhancer elements function in a complete or partially tissue-specific manner, that is, they will only enhance the transcription of genes in specific target tissues frequently. Many enhancers have now been characterized that play key roles in the regulation of gene expression. A striking feature of the characterized enhancers is that they exhibit tissue specificity. The tissue specificity of the enhancers is fascinating, however, we have not understood the molecular basis of this specificity. Presumably, the tissue specificity may result from interactions of enhancer sequences with transcriptional activators present only in cells in which a given set of genes is expressed. The ability of enhancers and silencers to act at a distance of 1000 bp or more is intriguing. At present this question cannot be answered in detail. But one point is clear, the factor bound at the enhancer sequence and the promoter sequence can act in a cooperative manner, either positively or negatively.

SPECIFIC TRANSCRIPTION FACTORS

Eukaryotic transcription begins with the formation of a pre-initiation complex formed by the amalgamation of a group

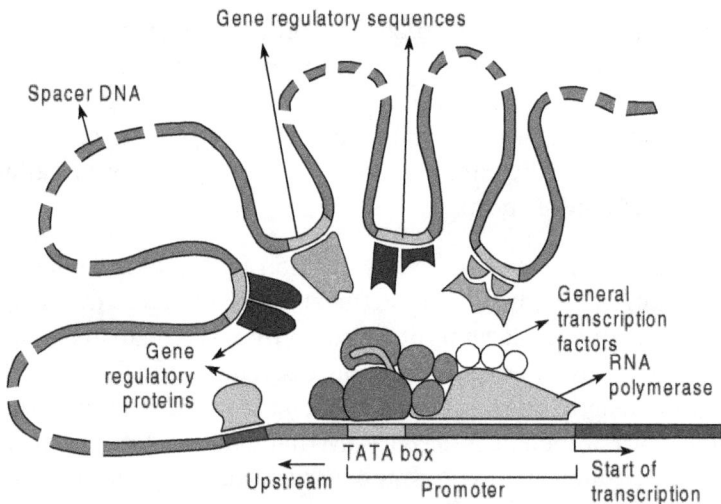

Figure 7.6 Molecular structure of transcription initiation factor and their interaction with enhancer sequence as well as protein

of general transcription factors (Figure 7.6). Proteins that exert control over transcription at specific promoters are the specific transcription factors. These proteins generally have two domains, a domain that recognizes a specific DNA sequence and a domain that recognizes another protein like the pre-initiation complex. A majority of specific transcription factors act by recruiting the components of the RNA polymerase holoenzyme. Thus, the binding of a specific transcription factor at a promoter is the first step in the formation of a pre-initiation complex at the promoter of a gene, e.g. dorsal wing development in drosophila.

Methylation of DNA

A small percentage of cystine residues are methylated in many eukaryotic organisms, mainly in CPG sequences. The degree of methylation of DNA is related to silencing of a gene. Genes that are dormant at one stage of development but active in another are usually less methylated. The site of addition of methyl group on the cysteine are shown in Figure 7.7.

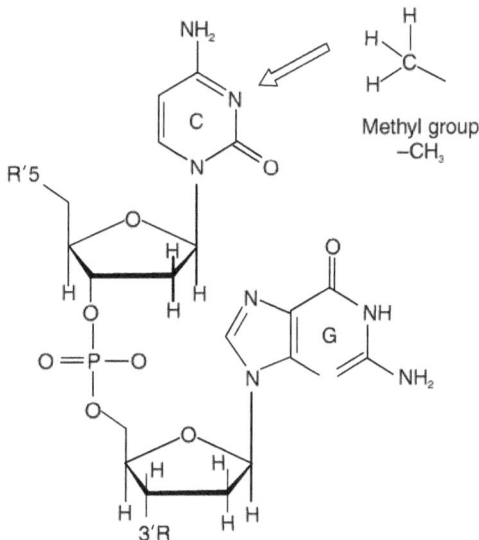

Figure 7.7   Sites of addition of methyl group on the cysteine

The essential function of m⁵cyt is to modify protein–DNA interaction. Methylation itself may not prevent transcription, but rather may be a signal for transcriptional inactivity. Further interest has been generated in the role of methylation in controlling gene expression by the discovery of Z-DNA and the fact that Z-DNA can be stabilized by methylation. This observation has led to a model of transcriptional regulation based on alternative DNA structure sequence that could exist as B-DNA while being transcribed. If the gene is to be silenced, (turned off) the CPG sequences are converted to stable Z-DNA by methylation which then blocks transcription. Methylation can exist in three states, methylated, half-methylated and fully methylated. After DNA replication, a half-methylated site is quickly converted to a fully methylated site. Symmetrical methylation occurs on both strands with the help of methylase enzyme acting only on a half-methylated site. Methylated sites would remain methylated, unmethylated sites would remain unmethylated. The complex secondary and tertiary folding of eukaryotic chromosomes probably involves specific protein–DNA interactions, which could be influenced by the distribution of m⁵cyt in the DNA. The high methylation of centromeric regions suggests a possible role in the mitosis and chromosome sorting.

REGULATION BY ALTERNATE TRANSCRIPT SPLICING

Transcriptional regulation plays an important role in controlling development in eukaryotes. The regulation of transcription process is well-documented in many systems and occurs in several ways. Regulation occurs by changing transcript stability, by differential transport to the cytoplasm and by differential translation of processed transcripts. One of the most spectacular examples of alternative modes of transcript splicing occurs in the case of the tropomyosin genes of drosophila and vertebrates.

Tropomyosins are a family of closely related proteins that mediate the interactions between actin and troponin and thus help regulate muscle contractions. Different tissues, both muscle and non-muscle, are characterized by the presence of different tropomyosin isoforms. It turns out that many of these isoforms are produced from the same gene by alternative splicing.

GENE REGULATION BY HORMONAL ACTION

Intercellular communication is a very important phenomenon in higher plants and animals. Signals originating in various glands and/or secretary cells somehow stimulate target tissue or target cells to undergo dramatic changes in their metabolic patterns. These changes frequently include altered patterns of differentiation. Peptide hormones, e.g. insulin, are large and hence do not enter into the cell. They show their effect by binding to cell surface receptors, which in turn activates the protein/enzyme transcription factors by phosphorylation mechanism. On the other hand, steroid hormones such as estrogen are small molecules, which readily enter through the plasma membrane. Once inside the target cell, they get tightly bound to specific receptor proteins, which are present only in the cytoplasm of the target cells.

The hormone–receptor protein complexes activate the transcription of specific genes in two possible ways. First, the hormone–receptor protein complexes interact with specific non-histone chromosomal protein and this interaction stimulates the transcription of corrector protein complexes and this interaction stimulates the transcription of correct genes. The hormone–receptor protein complexes activate transcription of target genes by binding to specific DNA sequences present in the *cis*-acting regulatory regions of genes.

Another hypothesis is that the hormone–receptor protein complexes interact with specific non-histone chromosomal proteins rather than directly with DNA. This interaction would then supposedly stimulate the transcription of the correct genes. It seems likely that the non-histone chromosomal proteins play an important role in the regulation of gene expression in eukaryotes. At present one cannot exclude the possibility that histone modifications or non-histone chromosomal proteins are involved in some aspects of hormone-regulated gene expression. On the other hand, the evidence that is available to date strongly suggests that hormone–receptor protein complexes activate gene expression by interacting directly with specific DNA sequences present within the enhancer or promoter regions that regulate the transcription of the target genes.

MECHANISMS OF GENE REGULATION IN EUKARYOTES

In the absence of precise information about the mechanisms that regulate gene expression in eukaryotes, many models were proposed. One of the more popular early models known as Britten–Davidson model or gene battery model was that given by R.J. Britten and E.H. Davidson in 1969. This model even though widely accepted, is only a theoretical model and lacks sound practical proof. The model predicts the presence of four types of sequences.

Producer gene   It is comparable to a structural gene in prokaryotes. It produces pre-mRNA, which after processing becomes mRNA. Its expression is under the control of many receptor sites.

Receptor site (gene)   It is comparable to the operator in bacterial operon. At least one such receptor site is assumed to be present adjacent to each producer gene. A specific receptor site is activated when a specific activator RNA or an activator protein, a product of integrator gene, complexes with it.

Integrator gene   Integrator gene is comparable to regulator gene and is responsible for the synthesis of an activator RNA molecule that may not give rise to proteins before it activates the receptor site. At least one integrator gene is present adjacent to each sensor site.

Sensor site   A sensor site regulates activity of an integrator gene which can be transcribed only when the sensor site is activated. The sensor sites are also regulatory sequences that are recognized by external stimuli, e.g. hormones, temperature.

According to the Britten–Davidson model, specific sensor genes represent sequence-specific binding sites (similar to CAP–cAMP binding site in the *E. coli*) that respond to a specific signal. When sensor genes receive the appropriate signals, they activate the transcription of the adjacent integrator genes. The integrator gene products will then interact in a sequence-specific manner with receptor genes. Britten and Davidson proposed that the integrator gene products are activator RNAs that interact directly with the receptor genes to trigger the transcription of the continuous producer genes.

It is also proposed that receptor sites and integrator genes may be repeated a number of times so as to control the activity of a large number of genes in the same cell. Repetition of receptor ensures that the same activator recognizes all of them and in this way several enzymes of one metabolic pathway are simultaneously synthesized. Transcription of the same gene may be needed in different developmental stages. This is achieved by the multiplicity of receptor sites and integrator genes. Each producer gene may have several receptor sites, each responding to one activator. Thus, though a single activator can recognize several genes, different activators may activate the same gene at different times. A set of structural genes controlled by one sensor site is termed as a battery. Sometimes when major changes are needed, it is necessary to activate several sets of genes. If one sensor site is associated with several integrators, it may cause transcription of all integrators simultaneously thus causing transcription of several producer genes through receptor sites. The repetition of integrator genes and receptor sites is consistent with the reports that state that sufficient repeated DNA occurs in the eukaryotic cells. The most attractive features of the Britten and Davidson model is that it provides a plausible reason for the observed pattern of interspersion of moderately repetitive DNA sequences and single copy DNA sequences. Direct evidence indicates that most structural genes are indeed single copy DNA sequences. The adjacent moderately repetitive DNA sequences would contain the various kinds of regulator genes (sensor, integrator and receptor genes).

REVIEW QUESTIONS

1. Describe the positive regulation with an example.

2. Explain in detail about *lac* operon.

3. Explain in detail about *trp* operon.

4. Write in detail about control of *trp* operon by attenuation.

5. Explain the mechanism of gene regulation in eukaryotes.

6. Explain the mechanism of gene regulation by *cis*-acting element.

7. Describe in detail about gene–battery model of gene regulation.

8. Write short notes on the following.

    i. Inducer

    ii. Repressor

  iii. Positive regulation

  iv. Negative regulation

   v. Activator

  vi. Repressor

  vii. Structural gene

 viii. Control gene

  ix. Regulator gene

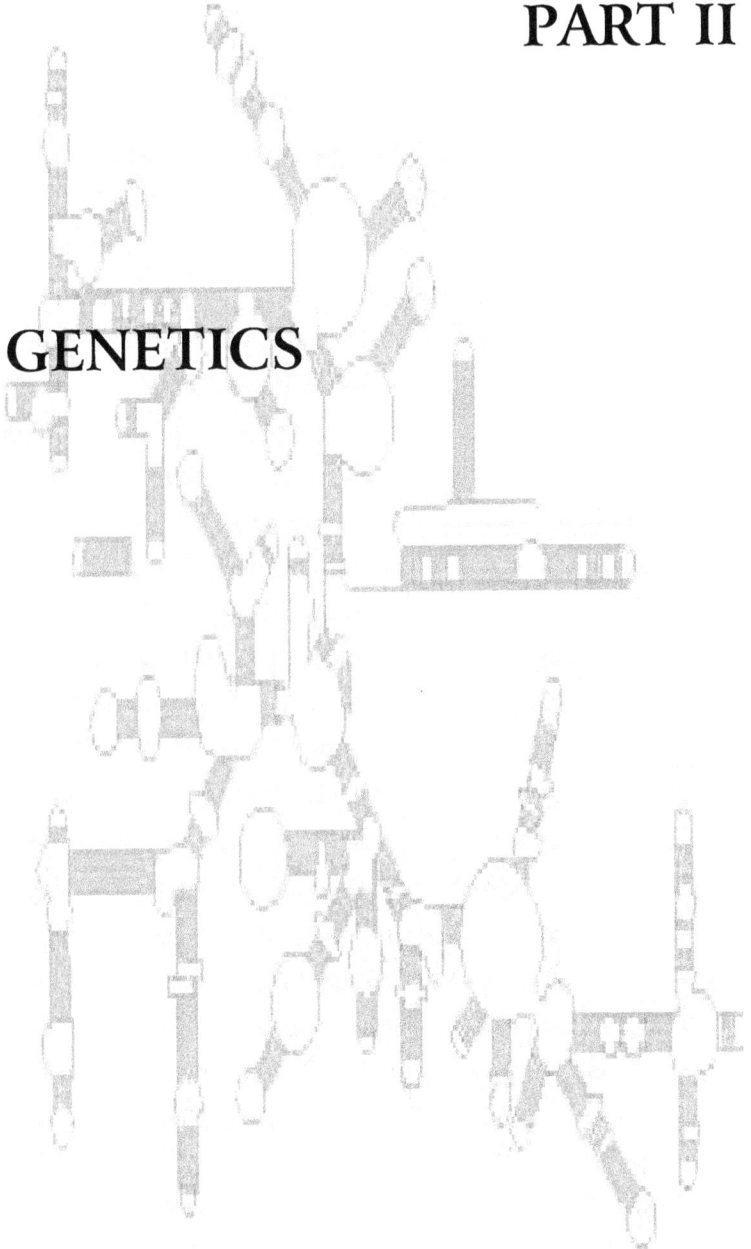

PART II

GENETICS

8

MENDEL'S LAWS

INTRODUCTION

For centuries, people realized that individual characteristics were passed on from parent to offspring. When a child is born people look for its resemblance with the parents or close blood relatives. Questions relating to the nature and the basis for this relationship have occupied the thoughts of man for centuries. However the mechanism of inheritance was not generally understood until early this century. Serious systematic attempts to seek answers to these questions began only in the 18th century. A number of scientists had worked on plant hybridization during the 18th and 19th centuries prior to Mendel. Koelreuter conducted extensive studies on hybridization and proposed the theory of uniformity and heterosis in F_1 and variations in F_2.

The following important conclusions were available to Mendel from the studies of his predecessors.

1. In F_1 hybrids, some characteristics are identical to those of one of the two parents, some are similar to those of

the other parent, while some others are intermediate between those of the two parents.

2. Characteristics of F_1 and F_2 progeny produced by reciprocal crosses are identical. This observation clearly demonstrates that the contributions of male and female parents to the characteristics of the progeny are equal.

3. F_1 progeny from a single cross are uniform in their characteristics. That is, all the plants in F_1 from a cross are similar to each other. But F_2 generation shows a large variation for different characteristics.

4. In F_2 generation, some plants have characteristics similar to one parent, while some others are similar to the other parent in their appearance. The appearance of parental forms in F_2 was called reversion. But a majority of the plants were intermediate in appearance between the two parents.

5. Some plants in F_2 have entirely new characteristic forms.

REASONS FOR FAILURE OF MENDEL'S PREDECESSORS

In his 1865 paper, Mendel presented a brilliant analysis of the deficiencies in the experimental approaches of his predecessors. These are summarized below.

1. The scientists studied the plant as a whole, i.e., its total appearance consisting of a large number of characters.

2. Therefore, the plants could not be classified into few clear-cut classes. These workers did not attempt an exhaustive classification of the different forms of the characteristics present in the progeny.

3. The scientists were more concerned with the description of various forms appearing in the progeny. An attempt to determine the frequencies of different characteristic forms in the progeny was not made.

4. In many cases, the data from different generations were not kept accurately and separately.

5. In many cases, a complete control on pollination in the F_1 was lacking.

6. In many studies, the F_1 was an interspecific hybrid exhibiting partial to considerable sterility.

7. The number of plants studied in F_2 was relatively small.

8. In addition, most of the characters studied by the earlier workers were quantitative in nature.

MENDEL

Gregor Johann Mendel was born on July 22, 1822 to peasant parents in a small agrarian town in Czechoslovakia. During his childhood he worked as a gardener, and as a young man attended the Olmutz Philosophical Institute. In 1843 he entered an Augustinian monastery in Brunn, Czechoslovakia. He was later sent to the University of Vienna to study. Mendel was inspired to study variance in plants by his professors at the University and his colleagues at the monastery. Between 1856 and 1863 Mendel cultivated and tested some 28,000 pea plants. His experiments brought forth two generalizations which later became known as Mendel's Laws of Heredity. Ironically, when Mendel's paper was published on 1866, it had little impact. It was not until the early 20th century that the enormity of his ideas was realized. It was not an accident that Mendel was able to explain the phenomenon of inheritance, a subject that had puzzled many previous researchers. His success was based on at at least three factors.

1. He had thoroughly documented and quantified his results.

2. He had concentrated initially on a single trait at a time, rather than on a number of traits which many of his contemporaries had tried to do.

3. The experimental organism he had selected, the garden pea, was well-suited for genetic investigations for several reasons.

The choice of pea for hybridization by Mendel was based on a deep understanding of the problems of such studies. Pea offered the following advantages as an experimental material.

1. In the available varieties, several characteristics had two contrasting forms, which were easily distinguishable from each other (Table 8.1). This permitted an easy classification of F_2 and F_3 progeny from various crosses into clear-cut classes.

2. The flower structure of pea ensures self-pollination. Mendel experimentally verified this.

3. Pea flowers are relatively large. Therefore, emasculation and pollination of pea flowers is quite easy.

4. The duration of pea crop is of a single season. As a result, every year one generation of pea can be grown.

5. Pea seeds are large and present no problem in germination. Pea plants are relatively easy to grow and each plant occupies only a small space.

6. In addition to pea, Mendel worked on rajma (*Phaseolus vulgaris* L.) as well. The results from the experiments on rajma were reported along with those on pea in the same paper.

Mendel's huge contributions to the world of science are in the form of three laws, which are widely called as Mendel's Laws of Inheritance. His work has stood the test of time, even as the discovery and understanding of chromosomes and genes has developed in the 140 years after he published his findings. New discoveries have found "exceptions" to Mendel's basic laws, but none of Mendel's findings have been proven to be totally wrong.

TERMINOLOGY AND DEFINITIONS

Alleles   Alternative forms of the same gene. Alleles for a trait are located at corresponding positions on homologous chromosomes. For example, there is a gene for hair texture (whether hair is curly or straight). One form of the hair texture gene codes for curly hair. A different code for the same gene makes hair straight. So the gene for hair texture exists as two alleles—one curly code, and one straight code. We could use a "C" for the curly allele, and a "c" for the straight allele. A person's genotype with respect to hair texture has three possibilities: CC, Cc, or cc. Thus, homozygous means having two of the same allele in the genotype (CC or cc). Heterozygous means one of each allele in the genotype (Cc).

Allelic pair   The combination of two alleles which comprise the gene pair.

Homozygote   An individual which contains only one allele at the allelic pair; for example DD is homozygous dominant and dd is homozygous recessive; pure lines are homozygous for the gene or character of interest.

Heterozygote   An individual that contains one of each member of the gene pair; for example the Dd heterozygote.

Back cross The cross of an F_1 hybrid to one of the homozygous parents; for pea plant height the cross would be Dd × DD or Dd × dd.

Test cross The cross of any individual to a homozygous recessive parent and is used to determine if the individual is homozygous dominant or heterozygous.

Monohybrid cross A cross between parents that differ at a single gene pair (usually AA × aa).

Monohybrid The offspring of two parents that are homozygous for alternate alleles of a gene pair. A monohybrid cross is not the cross of two monohybrids. Monohybrids are good for describing the relationship between alleles. When an allele is homozygous, it will show its phenotype. It is the phenotype of the heterozygote which permits us to determine the relationship of the alleles.

Dominance The ability of one allele to express its phenotype at the expense of an alternate allele. Generally the dominant allele will make a gene product that the recessive cannot; therefore the dominant allele will express itself whenever it is present.

Phenotype Physical appearance of a particular trait. Literally means "the form that is shown "outward".

Dominant The allele that expresses itself at the expense of an alternate allele; the phenotype that is expressed in the F_1 generation from the cross of two pure lines.

Recessive An allele whose expression is suppressed in the presence of a dominant allele; the phenotype that disappears in the F_1 generation from the cross of two pure lines and reappears in the F_2 generation.

Genotype The genes present in the DNA of an organism. We will use a pair of letters (Tt, YY, ss, etc.) to represent genotypes for one particular trait. There are always two letters in the genotype because (as a result of sexual reproduction) one code for the trait comes from female parent and the other comes from male parent, so every offspring gets two codes (two letters). But there are three possible genotypes—two big letters (like "TT"), one of each ("Tt"), or two lowercase letters ("tt"). Each possible combination has a term for it. When we have two capital or two lowercase letters in the genotype (TT or tt) it is called homozygous ("homo" means "the same"). Sometimes the term "pure" is used instead of homozygous. When the genotype is made up of one capital letter and one lowercase letter (Tt) it is called heterozygous.

Genotype = Genes present in an organism

TT = Homozygous = pure

Tt = Heterozygous = hybrid

tt = Homozygous = pure

Filial generation The generation produced by crossing two parents differing for one or more characters and later generations produced by selfing.

F_1 The generation produced by crossing two parents that differ for one or more characters.

F_2 The generation produced by selfing F_1 generation.

MENDEL'S LAWS

The three important laws of Mendel include;

1. The Law of Dominance
2. The Law of Segregation
3. The Law of Independent Assortment

The Law of Dominance

The first law of Mendel states that "In a cross of parents that are pure for contrasting traits, only one form of the trait will appear in the next generation. Offspring that are hybrid for a trait will have only the dominant trait in the phenotype."

While Mendel was crossing (reproducing) his pea plants (over and over and over again), he noticed something interesting. When he crossed pure tall plants with pure short plants, all the new pea plants (referred to as the F_1 generation) were tall. Similarly, crossing pure yellow-seeded pea plants and pure green-seeded pea plants produced an F_1 generation of all yellow seeded pea plants. The same was true for other pea traits. So, what he noticed was that when the parent plants had contrasting forms of a trait (tall vs short, green vs yellow, etc.) the phenotypes of the offspring resembled only one of the parent plants with respect to that trait (Table 8.1). So, Mendel proposed a law in which he proposed that "There is a factor that makes pea plants tall, and another factor that makes pea plants short. Furthermore, when the factors were mixed, the tall factor seemed to dominate over the short factor."

Table 8.1 Results of F_1 generation obtained from cross in pea plant

Parent pea plants	F_1 pea plants
tall stem × short stem	All tall stems
yellow seeds × green seeds	All yellow seeds
green pea pods × yellow pea pods	All green pea pods
round seeds × wrinkled seeds	All round seeds
axial flowers × terminal flowers	All axial flowers

Now, from our modern wisdom, we use "allele" or "gene" instead of what Mendel called "factors". There is a gene in the DNA of pea plants that controls plant height (makes them either tall or short). One form of the gene (allele) codes for tall, and the other allele for plant height codes for short. For abbreviations, we use the capital "T" for the dominant tall allele, and the lowercase "t" for the recessive short allele (Table 8.2).

Table 8.2 Genetic terminology

Genotype symbol	Genotype
TT	Homozygous dominant or pure tall
Tt	Heterozygous or hybrid
tt	Homozygous recessive or pure short

Confirmation of Mendel's First Law

With these observations, Mendel could form a hypothesis about segregation. To test this hypothesis, Mendel selfed the F_2 plants. If his law was correct he could predict what the results would be. And indeed, the results occurred as he expected.

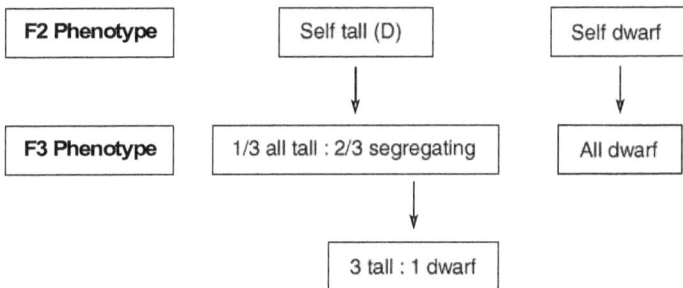

F2 Phenotype	Self tall (D)	Self dwarf
F3 Phenotype	1/3 all tall : 2/3 segregating	All dwarf
	3 tall : 1 dwarf	

From these results we can now confirm the genotype of the F_2 individuals.

Table 8.3   Genetic description of F_2 and F_3 generation offsprings

Phenotypes	Genotypes	Genetic description
F_2 Tall plants	1/3 DD	Pure line homozygote dominant
	2/3 Dd	Heterozygotes
F_2 Dwarf plants	All dd	Pure line homozygote recessive

Thus the F_2 is genotypically $\frac{1}{4}$ Dd : $\frac{1}{2}$ Dd : $\frac{1}{4}$ dd . This data was also available from the Punnett square using the gametes from the F_1 individual. So although the phenotypic ratio is 3:1 the genotypic ratio is 1:2:1 (Table 8.3).

Mendel performed one other cross to confirm the hypothesis of segregation—the backcross. We must remember that the first cross is between two pure line parents to produce an F_1 heterozygote.

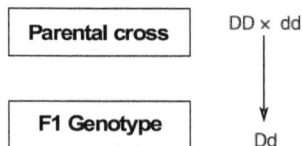

At this point instead of selfing the F_1, Mendel crossed it to a pure line, homozygote dwarf plant.

Mendel's Law of Segregation

The second law states that "during the formation of gametes (eggs or sperm), the two alleles responsible for a trait separate from each other. Alleles for a trait are then "recombined" at fertilization, producing the genotype for the traits of the offspring." So, he took two of the "F_1" generation (which are tall) and crossed them. His new batch of pea plants (the "F_2" generation) is about 3/4 tall and 1/4 short.

The parent plants for this cross each have one tall factor that dominates the short factor and causes them to grow tall. To get short plants from these parents, the tall and short factors must separate, otherwise a plant with just short factors could not be produced. The factors must "segregate" themselves somewhere between the production of sex cells and fertilization.

It can be seen from the p-square that any time two hybrids are crossed, 3 of the 4 boxes will produce an organism with the dominant trait (in this example "TT", "Tt", and "Tt"), and 1 of the 4 boxes ends up homozygous recessive, producing an organism with the recessive phenotype ("tt" in this example) (Table 8.4).

	T	t
T	TT × tt	Tt
t	Tt	tt

When two parents have the same phenotype for a trait but some of their offspring look different with respect to that trait, the parents must be hybrid for that trait.

Table 8.4   Genotype and phenotypes of F_1 and F_2 generation offsprings

| Parent pea plants | | Offspring | |
| (Two members of F₁ generation) | | (F₂ generation) | |
Genotypes	Phenotypes	Genotypes	Phenotypes
Tt × Tt	tall × tall	25% TT	75% tall
		50% Tt	25% short
		25% tt	

Law of Independent Assortment

The third law states that "Alleles for different traits are distributed to sex cells (offspring) independently of one another."

Mendel noticed during all his work that the height (tall or short), of the plant and the shape (round or wrinkled) of the seeds and the colour (green or yellow), of the pods had no impact on one another. In other words, being tall did not automatically mean that the plants had to have green pods, nor did green pods have to be filled only with wrinkled seeds, the different traits seem to be inherited independently. It involves what is known as a "dihybrid cross", meaning that the parents are hybrid for two different traits (Table 8.5).

The genotypes of the parent pea plants will be: RrGg × RrGg where

R = dominant allele for round seeds

r = recessive allele for wrinkled seeds

G = dominant allele for green pods

g = recessive allele for yellow pods

Thus we are dealing with two different traits: (1) seed texture (round or wrinkled) and (2) pod colour (green or yellow). Also each parent is hybrid for each trait (one dominant and one recessive allele for each trait).

Table 8.5 Punnett square showing all possible combinations of genotypes that could arise from self-crossing of F₁ generation

	RG	Rg	RG	rg
RG	RRGG (Round, Yellow)	RRGg (Round, Yellow)	RrGG (Round, Yellow)	RrGg (Round, Yellow)
Rg	RRGg (Round, Yellow)	RRgg (Round green)	RrGg (Round, Yellow)	Rrgg (Round green)
rG	RrGG (Round, Yellow)	RrGg (Round, Yellow)	rrGG (wrinkled Yellow)	rrGg (wrinkled Yellow)
rg	RrGg (Round, Yellow)	Rrgg (Round green)	rrGg (wrinkled Yellow)	rrgg (wrinkled green)

The results from a dihybrid cross are always the same:

9/16 boxes (offspring) show dominant phenotype for both traits (round and yellow), 3/16 show dominant phenotype for first trait and recessive for second (round and green), 3/16 show recessive phenotype for first trait and dominant form for second (wrinkled and yellow), 1/16 show recessive form of both traits (wrinkled and green).

So, as can be seen from the results, a green pod can have round or wrinkled seeds, and the same is true of a yellow pod. The different traits do not influence the inheritance of each other. They are inherited independently.

Interesting to note is that if we consider one trait at a time, we get "the usual" 3:1 ratio of a single hybrid cross (like we did for the Law of Segregation). For example, let us compare the colour trait in the offspring; 12 green and 4 yellow (3:1 dominant:recessive) and the same for the seed texture 12 round and 4 wrinkled (3:1 ratio). The traits are inherited independently of each other.

Summary of Mendel's Laws

Law	Parent Cross	Offspring
Dominance	TT × tt (Tall × Short)	100% Tt Tall
Segregation	Tt × Tt (Tall × Tall)	75% Tall 25% Short
Independent Assortment	RrGg × RrGg (Round and green) × (Round and green)	9/16 round seeds and green pods 3/16 round seeds and yellow pods 3/16 wrinkled seeds and green pods 1/16 wrinkled seeds and yellow pods

REVIEW QUESTIONS

1. Write in detail about the conclusion available to Mendel before his theory.

2. Describe the reason for failure of Mendel's predecessors.

3. Explain in detail the laws of Mendel.

4. Write short notes on the following.

    i. Allele

    ii. Dominance

    iii. Monohybrid cross

    iv. Dihybrid cross

    v. Genotype

    vi. Phenotype

    vii. Law of dominance

    viii. Law of independent assortment

    ix. F_1 generation

9

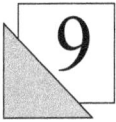

GENE INTERACTION

The elegance of Mendel's experiments was partly due to the complete consistency between his observations and the hypothesis he developed. Mendel was lucky that in his studies the characteristics he chose were governed by a single gene. However after Mendel's work was rediscovered, it became clear that simple Mendelian model is not sufficient to predict experimental observations in all situations. Two or more genes govern the expression of many characters in almost all organisms. These genes affect the development of the concerned characters in various ways.

The phenomena of two or more genes governing the development of a single character in such a way that they affect the expressions of each other in various ways is known as gene interaction. Gene interaction may involve two or more genes. But all the examples described below relate to interaction between two genes. Therefore, these cases are also referred to as the modification of the dihybrid ratio.

BATESON AND PUNNETT EXPERIMENT

A classical example of gene interaction based on the results of cross between different breeds of chicken was reported. In the early part of the 20th century, W. Bateson and R.C. Punnett crossed domestic breeds of chickens such as Wyandottes, brahmas, leghorns, etc., having different types of combs. Wyandottes have a characteristic type of comb called rose whereas brahmas have a pea comb and leghorns have a single comb. When a cross is done between Wyandottes and Brahmas, all F_1 chicken had combs with a phenotype not expressed in either parent. When the F_1 chicken were mated among themselves and large F_2 population were produced, a familiar dihybrid ratio 9:3:3:1 was recognized but the phenotype representing two of the four classes were different from those expressed in the parents. About 9/16 of the F_2 birds were walnut, 3/16 were rose, 3/16 were pea and 1/16 had single combs. Neither single comb nor walnut was expressed in the original parent lines. These two phenotypes were explained as the result of gene product interaction. Results indicated that two different allelic pairs were involved—one pair was introduced by the rose comb parent and one by the pea comb parent. A gene for rose and a gene for pea would interact and produce walnut as in F_1.

Analysis of the F_2 results and appropriate test crosses indicated that the 9/16 class with the two dominant genes (R-P) was walnut like the F_1 chicken. The 1/16 class representing the full receives combination (rrpp) was characterized as a single comb. The two 3/16 (rose and pea) classes were RR-pp and rr-pp. It was then determined that the homologous genotype of the rose-combed parent (Wyandotte) was RRpp and that of the pea-combed parent (brahmas) was rrpp. Although the usual 9:3:3:1 ratio was obtained, the result from this cross was unusual in two important respects.

1. The offspring obtained from a cross of Wyandottes and Brahmas showed a comb of walnut type, a characteristic which none of the parents have.

2. Two phenotypes (walnut and single) not expressed in the original appeared in F_2.

The genes R and P were non-allelic but each was a dominant over its allele. When R and P were together, as in the F_1 (RrPp), the two different products interacted to produce a walnut comb. The two non-allelic genes R and P acted independently in different ways similar to the ways in which dominant alleles act.

TYPES OF GENE INTERACTIONS

Dominance

Dominance is the interaction between alleles, and the dominant allele is usually the one that produces a functional product while recessive allele does not. For example if the phenotype of allele A is dominant, AA and Aa individuals are alike phenotypically. In the heterozygous condition (Aa), allele "a" is completely masked and the trait is recessive. Phenotypes attributable to single allele substitutions are called dominants and the phenomenon is called as dominance. Several different patterns of dominance are also documented.

Co-dominance   When both alleles of a pair are fully expressed in a heterozygote, they are called co-dominant alleles. Such alleles exhibit a unique pattern of expression with heterozygotes being phenotypically distinguishable from both of the homozygotes and expressing both alleles equally. In general, two products are the same with respect to function but different in exact amino acid sequence. A phenotypic ratio of 1:2:1 has thus replaced the

3:1 ratio because the alleles are both expressed in heterozygote, that is, the alleles are co-dominant.

The MN blood group antigens in humans are a good example. Allele L^M for M-type blood is co-dominant with its allele L^N for N-type blood. The heterozygote ($L^M L^N$) expresses the characteristics of both M and N antigens on their surfaces (MN-type blood).

Incomplete dominance   In incomplete dominance, alleles may produce the same product, but in lesser quantity as compared with the dominant allele. In heterozygous condition, the total product is intermediate between that of the dominant and recessive alleles. The phenotypic ratio is 1:2:1. For example, in snapdragons, the homozygous dominant for flower colour will show red colour and homozygous recessive will be white. In contrast, the heterozygous will have pink colour. This is against Mendel's theory of dominance, as a heterozygous plant should show the characteristic of dominant, i.e., red.

Lethal Genes

Genes which affect the viability of an organism are called as lethal genes and the phenomenon is called as lethality. If the lethal effect is dominant and immediate in expression, all individuals carrying the gene will die and the gene will be lost. Some dominant lethals, however, have a delayed effect so that the organism lives for a long time, e.g. Huntington's disease. Recessive lethals carried in the heterozygous condition have no effect but may come to expression when matings between carriers occur. The phenotypic ratio is 2:1. For example, in mice, yellow colour is dominant over agouti. But when a cross takes place between two yellow coloured mice, the result was 2 yellow and 1 agouti (2:1) instead of 3:1. This was because the yellow colour is a dominant lethal. The existing yellow coloured mice were in heterozygous conditions. All the yellow coloured mice in homozygous condition died at the embryo stage.

Pleiotrophy

The phenomenon in which a single gene affects two or more characteristics is called pleiotrophy. For example, in humans a rare genetic disorder, phenylketonuria, occurs in individuals homozygous for a defective recessive allele. These people lack the enzyme necessary for the normal metabolism of the amino acid phenylalanine. In addition, they possess smaller head and somewhat lighter hair.

Duplicate Genes

The presence of a single dominant allele of any one of the two genes governing the trait produces the dominant phenotype. The recessive phenotype is produced only when both the genes are in the homozygous recessive state, i.e., the dominant genes when present together or alone will produce the same phenotype, but when they are in recessive state together, they will produce different phenotypes. This gene interaction produces a ratio of 15:1.

For example, the non-floating habit in rice is controlled by two dominant genes Dw_1 and Dw_2. When a non-floating rice strain with the genotype (Dw_1, Dw_1, Dw_2, Dw_2) is crossed with a floating strain (dw_1, dw_1, dw_2, dw_2), the F_1 is non-floating. In the F_2 generation, on an average 9 plants will have at least one dominant allele of both the gene, 3 plants will have at least one dominant allele of one of the two genes (homozygous for other gene) and 3 other individuals will have one dominant allele of the other gene. All these 15 individuals will have non-floating habit, and only one of the 16 possible zygotic combinations will be homozygous recessive for both the genes, dw_1 and dw_2.

Complementary Genes

In this type of gene interaction, the production of one phenotype requires the presence of dominant alleles of both the genes controlling the character. When any one of the two or both the genes are present in the homozygous recessive state, the contrasting phenotype is produced. Thus any one of the two dominant genes is unable to produce the phenotype when it is alone. But the dominant alleles of the two genes complement each other to produce the concerned phenotype when they are together. This type of interaction shows a ratio of 9:7.

For example, in sweet pea, the development of purple coloured flowers requires the presence of two dominant genes C and B. When either C or R (e.g. CCrr, ccRR) or both of them (e.g. ccrr) are present in homozygous recessive condition, purple coloured flower cannot be produced. As a result, white flowers are obtained.

Supplementary Genes

In supplementary genes, the dominant allele of one gene produces a phenotypic effect. The dominant allele of the other gene does not produce any phenotypic effect on its own. But when it is present with the dominant allele of the first gene, it modifies the phenotypic effect produced by the first gene. That is, the dominant allele of one gene is necessary for the development of the concerned phenotype, while that of the other gene modifies the phenotypic expression of the first gene. This interaction shows a ratio of 9:3:4.

For example, the development of aleurone colour in maize is governed by two completely dominant genes R and P. A plant producing purple coloured grain (RRPP) is crossed with a plant producing white coloured grain (rrpp). The F_1 plants produce purple colour. In F_2, 9 out of the 16 zygotic combinations will have at least one dominant allele either R or P. They will develop red colour since the recessive allele has no effect on colour production. Three other zygotes will be homozygous rr but will have the dominant allele P. These seeds will be white since rr is unable to produce colour and P does not produce any colour. The remaining four zygotes will be homozygous recessive for both the genes (rrpp) and will produce white seed.

REVIEW QUESTIONS

1. Explain the phenomenon of dominance in detail.

2. Explain in detail any two types of gene interactions with examples.

3. Write short notes on the following.

    i. Co-dominace

    ii. Incomplete dominace

    iii. Lethal genes

    iv. Pleiotrophy

    v. Duplicate genes

10

LINKAGE AND CROSSING OVER

LINKAGE

After Sutton suggested the chromosomal theory of inheritance in 1903, evidence accumulated for the idea that genes were located on the chromosome. Morgan showed that there could be genes more than chromosomes. Sturtevant first showed that genes are arranged in a linear fashion on the chromosome. The concept of presence of genes on chromosomes is universally accepted. But, genes present on single chromosome move together to the same pole in every meiosis step. As a consequence such genes would fail to show independent segregation. But in real situation, when a test cross is carried in pea plant (flower colour and pollen shape), the result showed $7:1:1:7$ ratio instead of $1:1:1:1$ ratio as per Mendelian law of independent segregation. Bateson and Punnett observed these variations and were puzzled. Hence they named this peculiar behaviour as coupling and repulsion phase. But Morgan in 1911 proposed the new concept of linkage.

1. Genes are arranged on the chromosome in a linear fashion.

2. Genes located in the same chromosome tend to stay together during every generation or inheritance.

3. The distance between two genes decides whether they will be inherited together and if so with what frequency.

Morgan proposed finally, "the ability or tendency of two genes to stay together during inheritance is known as linkage. Coupling and repulsion phases are two aspects of linkage".

For example, let us consider two characteristics—red coloured flowers and long leaves—of a plant. The seeds from such plant are grown. If the plant with long leaf produces red coloured flower, the seed of F_1 generations are grown again. The results are compared, i.e., whether the plant with long leaves is producing red coloured flowers or not. If these two characteristics are present simultaneously together in every generation, then these genes are said to be "linked", and this phenomenon or transmission pattern of the linked genes is called "linkage."

Two or more genes may stay together (during inheritance) because they are located in the same chromosome. Linked genes do not show independent segregation. Hence the frequencies of parental character combinations are markedly more than expected, while those of recombinants or new character combinations are considerably low.

Linkage is further classified into two groups, complete or incomplete linkage depending upon the absence (complete) or presence of recombinant types in the test cross progeny. For example, if genes L and G (long wings and grey bodies) are present in one chromosome, while their recessive alleles l and g (short wing and black bodies) are on the homologous chromosome. Each chromosome behaves as a unit during cell division. Therefore, gene L and G would move to one pole, while l and g would move to the opposite pole.

If this always happened the F$_1$ (LlGg) would produce only two types of gametes (i.e., LLGG long wings, grey bodies and llgg short wings, black bodies) in test progeny. When only combinations of parental characteristics are recovered in test cross progeny, it is called complete linkage.

However, linked genes do not always move to the same pole. Sometimes their alleles recombine to produce recombinant gamete Lg and lg. This yields the recombinant phenotype long winged black bodies (LLgg) and short winged grey bodies (llGG). Here the recombinant types are also recovered in the test cross progeny and are called incomplete linkage.

Linkage is also classified as coupling and repulsion phase linkage. If dominant alleles of the linked genes are present in the same chromosome, then it is referred to as coupling phase. In contrast, in repulsion phase linkage, the dominant allele of one gene is present with the recessive allele of the other gene in the same chromosome.

Coupling Phase

In drosophila, a dominant L produces long wings, while its recessive allele, l produces short wings. Another dominant gene G governs grey bodies, whereas its recessive allele governs black bodies. When long-winged grey coloured Drosophila (LLGG) were crossed with short-winged black coloured (llgg) ones, F$_1$ progeny was long-winged and grey coloured. The F$_1$ (llgg) was test crossed with the double recessive drosophila (llgg). The result showed that 82% of progeny show parental combination. While the 18% showed recombinant progeny properties (i.e., long winged black coloured or short winged grey coloured). Clearly, the four phenotypic classes are not present in the expected ratio of 1:1:1:1. The phenotypic classes long-winged grey coloured and short-winged black coloured have a much higher frequency than the expected 25%. These two character combinations are referred to as parental combination or parental phenotypes or parental types because they have the same character combinations that were present in the two parents of F_1. The remaining two phenotypic classes, coloured shrunken and colourless full are far less frequent than the expected 25%. These two character combinations are called recombinant phenotypes or recombinant types, since they are obtained by reshuffling of the characters present in the two parents of the F_1. From the results, it is seen that dominant genes L and G have strong attractions for each other. This phase is called as coupling phase or *cis*-configuration.

In other words, coupling phase is one in which two dominant genes are linked and they do not get separated or transferred on to other chromosomes by recombination.

Repulsion Phase

When long-winged black bodies (LLDD) are crossed with short-winged grey bodies (llBB), the F_1 were long-winged and grey-bodied (LlGg). But when the F_1 were test crossed with the double recessive strain (llgg), the parental type long winged black bodies and short winged, grey bodies were more frequent than the recombinant types. It appears as if in this cross, the dominant genes did not like one another, hence they got separated. This situation is referred to as repulsion phase or *trans*-configuration.

Coupling and repulsion phases are obviously two phenomena of linkage. The only point to consider is whether two dominant genes or characters are existing simultaneously on one chromosome or not. In other words, if the parental combination continues to exist in F_1, F_2 and in test cross, i.e., in every generation, then such combination of linkage is called complete linkage. On the other hand, if the parental combination continues to exist only in every alternate generation, i.e., F_1, F_3, F_5 ...... F_{n-1}, then such combination of linkage is called as incomplete linkage.

Linkage between two dominant genes produces significant deviation from the typical dihybrid ratio $1:1:1:1$ (test cross) and $9:3:3:1$ (in F_2 generation). This deviation is the most easily detected in test cross. In the case of linkage, the two parental types are most frequent. In the linkage group, the percentage of parental types will be more than 50% and the recombinant percentage type will be less than 50%. It may be seen that the two parental types have comparable frequencies. Similarly the frequencies of the two recombinant types are also comparable. This type of test cross data is a sure indication of linkage. In fact, this relationship can be used as a safe guide to identify the two types in the test cross data whenever the identity of the parental type is not known. All the genes present on single chromosomes are grouped as one linkage group. Thus, the number of linkage groups present in a species is equal to the number of chromosomes present in that species. For example, the number of linkage groups in humans is 23 and that of maize is 10.

Genes present in one linkage group can be represented on a single straight line in the same order in which they are present in the chromosome, and the distance between two linked genes is proportional to the frequency of recombination between them which can be depicted in a diagram. Such a diagram with the linear order and recombination frequencies depicted is called as linkage map or genetic map or chromosome map. Before preparing a chromosome map, the sequence of genes in the chromosome and frequencies of recombination between linked genes must be known. An appropriate test cross will determine the recombination frequencies between linked genes. Each recombination frequency is used as a map unit for preparing the linkage map. A map unit is defined as that distance between genes for which one product of meiosis out of 100 is recombinant or the recombination frequency between two genes is one. Then the two genes are separated or present on the chromosome at a distance of a map unit. Map unit is an imaginary distance and does not represent the actual physical distance between two linked genes in the chromosomes. Sometimes a map unit is referred to as centimorgan in the honour of Thomas Hunt Morgan. The sequence of linked genes is determined by studying the test cross for three linked genes at a time. The data from such a test cross provides information on the order of the three genes in the chromosome as well as the frequency of recombination among them. To begin with, three linked genes are mapped subsequently, a three-point test cross involving only two of the three already mapped linked genes and a new gene expected to be linked with them is studied to map the new gene. It is desirable to include only those genes that show less than 20% recombination with each other.

Symbols The linked genes are conventionally symbolized by listing to the left of a slash the alleles on one homologue and to the right of the slash the alleles on the other homologue. The genes are listed in the same order on both sides of the slash, e.g. AB/ab or Ab/aB.

CROSSING OVER

In linkage analysis, two crosses are described, regardless of whether the cross was done during coupling or repulsion. The F_1 produces gametes, 18% of which carried recombination progeny.

The reason for the production of the recombinant progeny with a fixed frequency is due to the occurrence of a phenomenon called recombination or crossing over. Crossing over is the exchange of strictly homologous segments between non-sister chromatids of homologous chromosomes. Crossing over occurs during pachytene stage of meiosis-1 and is responsible for recombination between linked genes.

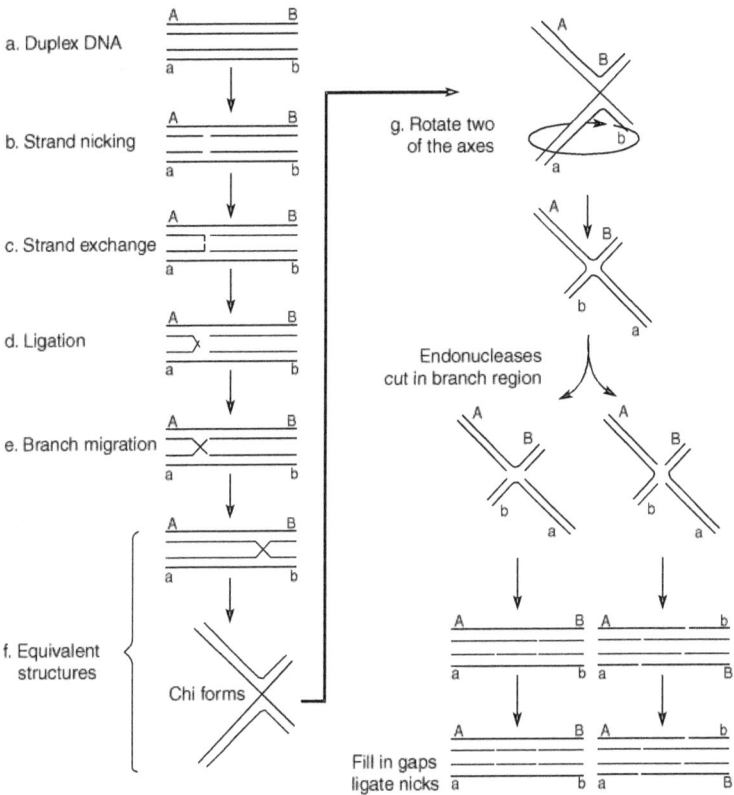

Figure 10.1   Mechanism of crossing over

During pachytene stage, each chromosome of a bivalent (chromosome pair) has two chromatids. Thus each bivalent contains four chromatids or strands (four-strand stage). Generally, one chromatid from each homologue is involved in crossing over. In the crossing over a segment of one chromatid becomes attached in place of the homologous segment of the non-sister chromatid and vice versa. Crossing over involves the breakage of each of two homologous chromosomes at precisely homologous points in the two non-sister chromatids and then exchange of parts. This produces an X-like structure at the point of exchange of the chromatid segments. This structure is called chiasma (plural chaismata). Each homologous pair can show one to several chaismata. Chiasmata can be anywhere along the length of a pair but normally it is seen in different positions in different meiocytes. Chaismata occur more or less randomly. The mechanism of crossing over is shown in Figure 10.1.

Crossing over occurs at the pachytene stage after the synapsis of the homologous chromosomes has occurred in prophase-I of meiosis. Since chromosome replication occurs during interphase, meiotic crossing over occurs in the post-replication tetrad stage, that is, after each chromosome has doubled such that four chromatids are present for each pair of homologous chromosomes.

Each event of crossing over produces two recombinant chromatids called as crossover-chromatids and two original chromosomes (i.e., which have not undergone any crossing over) are called as non-crossover chromatids. The crossover chromatids will have new combinations of the linked genes, i.e., will be recombinant. Gametes carrying them will produce the recombinant phenotype in test. Hence, these phenotypes are called as crossover types, whereas the other chromatids will give rise to the parental type character and are therefore called as parental phenotypes or non-crossover types. Genes on one chromosome are said to be linked for the obvious reason that they are physically linked or joined together by the segment of chromosome between them. The position of alleles of one gene on a chromosome is always the same and fixed. The location of a gene on a chromosome is called a locus (plural loci). Thus, we can say that two genes whose loci are on the same chromosome are said to be linked. If both the genes are dominant or both are recessive, then such an arrangement is called as *cis*-hybrid. But if one gene is dominant and other is recessive and if both the genes are linked, then it is called as *trans*-hybrid.

Chromosomes with recombinant combinations of linked genes are formed by the occurrence of crossing over in the region between the two loci. The probability that crossing over will occur between two loci increases with increasing distance between the two genes. The frequency of crossing over between two genes can be estimated as the frequency of recombinant progeny obtained in a test cross for these genes. This frequency is usually expressed as a percentage.

$$\frac{\text{Frequency of}}{\text{crossing over}} = \frac{\text{Number of recombinants}}{\text{Total number of progeny in test cross}} \times 100$$

Cytological Basis of Crossing Over

Morgan first proposed crossing over to explain the formation of recombinant combinations of genes that were shown to be linked by genetic data. He proposed that this linkage was the result of the location of these genes on the same chromosomes. If crossing over occurs, one might expect to be able to observe it under the microscope. In fact crossing over was first detected in amphibians by F. Jannsens. Stern and Creinghton, and McClintock are two classic works in genetics. They provided confirmation of Morgan's hypothesis that crossing over involves the interchange of parts of homologous chromosomes, they also provided strong evidence that indicated that genes are indeed located on the chromosome.

Direct cytological evidence that homologous chromosomes exchange parts during crossing over was first obtained in 1931 by C. Stern. Normally, the two chromosomes of any homologous pair are morphologically indistinguishable, that is they were not entirely homologous. The chromosome pairs studied by these workers were homologous along most of their length, such that they paired and segregated normally during meiosis. The homologues differed at their end having distinct morphological features that could be recognized by microscopy.

Stern studied two X chromosomes that differed from the normal X chromosome of *Drosophila*. One X chromosome had a part of a Y chromosome attached to one end. The second X chromosome was shorter than the normal X chromosome. This chromosome has a recessive gene car (produces carnation eye colour) and a dominant gene B (Bar eye shape), whereas the first chromosome had the dominant gene car^+ (produces red colour) and the recessive gene B^+ (round eye shape). Crosses were done to produce female flies heterozygous for these two morphologically distinguishable X chromosomes. These heterozygous females with *cis* configuration were crossed to males with car/B^+ (carnation eye colour, round eye shaped). As expected, the following four types of flies were recovered in the test cross progeny car^+/B^+ (red eye colour, round eye shape), car/B (carnation eye colour, bar eye shape), car/B^+ (carnation eye colour, round eye shape) and car^+/B (red eye colour, bar eye shape). As two out of these four phenotypes, carnation individuals are expected to carry one short X chromosome, while the red, normal flies would have one long X chromosome with an attached Y segment, whereas the two phenotypes, viz. red bar and a carnation, normal are crossovers. Hence, carnation, normal (car/B) flies are expected to have a normal or long X chromosome without the attached Y segment. In contrast red, bar (car^+/B) individuals will have one short X chromosome with the attached Y segment.

Stern determined the genotypes of the progeny both by chi-square method and by cytological observations. The results were precisely those predicted. Therefore he concluded that during meiosis, there is exchange of precisely homologous chromatin segments between homologous chromosomes (crossing over) and that crossing over is responsible for recombination between linked genes.

Barbara McClintock and Harriet Creighton provided a direct physical demonstration of recombination in 1931. By examining maize chromosomes microscopically, they could detect recombinations between two easily identifiable features of a particular chromosome both by physical observation as well as genetic recombinants. Creighton and McClintock worked with a strain of maize which contains an abnormal chromosome 9, especially at the end. One end had a knob and the other had an added piece of chromatin from another chromosome. This knobbed chromosome was thus clearly different from its normal homologue. It also connects the dominant coloured (c) allele and the recessive waxy texture (wx) allele. After mapping studies showed that c was very close to the knob and wx was close to the added piece of chromatin. Dihybrid plant with heteromorphic chromosome was crossed with the normal homomorphic plant (colourless and non-waxy phenotype). If a crossover occurred during meiosis in the dihybrid in the region between c and wx, a physical crossover, visible cytological by should also occur causing the knob to become associated with an otherwise normal chromosome and the extra piece of chromosome 9 to be associated with a knobless chromosome. Four types of gametes would result. Out of the 8 offspring examined, one offspring had coloured waxy phenotype, all had a knobbed interchange chromosome as well as a normal homologue. Those with the colourless, waxy phenotype had a knobless interchange chromosome. All of the coloured, nonwaxy phenotypes had a knobbed normal chromosome. Of those that contained only normal chromosomes, some were wxwx and some were heterozygotes. Of those containing interchange chromosomes, two were heterozygous and two were homozygous. These represent a crossover in the region between the waxy locus and the extra piece and chromatin, producing a knobless – c – wx extra piece chromosome.

Crieghton and McClintock concluded that pairing chromosomes, heteromorphic in two regions, have been shown to exchange parents at the same time they exchange genes assigned to these regions.

REVIEW QUESTIONS

1. Write in detail about crossing over.
2. Explain the phenomenon of linkage.
3. Enumerate the cytological basic of crossing over.
4. Write short notes on the following.
    i. Sutton's chromosome theory
    ii. Coupling phase
    iii. Repulsion phase
    iv. Morgan's concept of linkage
    v. Complete linkage
    vi. Incomplete linkage

11

MUTATIONS

Mutation is defined as the sudden heritable change in the genetic material of an organism. The term mutation is applicable to both the change in genetic material and to the process by which the change occurs. Thus the term mutation is used to define the process as well as the effect. Mutation is simply an alteration in the nucleotide sequence of a DNA molecule. Physical agents like UV or chemical molecules like Et.Br can cause mutations. Molecules or agents that cause mutations are called as mutagens. Mutations occur randomly or non-specifically and there is no defined process or machinery to carry out mutation in a cell or organism. Recombination on the other hand occurs at a particular time, with the help of a set of enzymes (e.g. Rec) and in a defined process. Thus mutation and recombination are not the same. But mutation and recombination are central events in genetics and evolution.

The process by which the rate of mutation is increased either by using physical or chemical agents is called as mutagenesis. Mutations created in an individual by the process of mutagenesis are called as induced mutations. Those mutations that occur without any cause are called as spontaneous mutations. Spontaneous mutations are regarded as accidental, unintended, unidirectional, unoriented random events. Even today we have no way to tell where or when a mutation will occur in a cell, cell organelle or gene. Spontaneous mutations are deviations from a natural process. Literally speaking, it is not possible to prove that a particular mutation is a spontaneous mutation or an induced mutation. But for the sake of convenience, mutations are classified into various groups and subgroups. Mutations can be looked at from three different angles.

1. DNA mutations

2. Chromosomal mutations

3. Organism or phenotypic mutations

All mutations take place in the gene, DNA sequence or nucleotide, but depending upon the site where their effect is visualized, they are classified into the types mentioned above.

DNA MUTATIONS

DNA mutations are those which occur or change a DNA sequence. These mutations are further classified into three main categories based upon whether they alter

❑   the DNA sequence

❑   the function of the gene

❑   the expression of the DNA sequence (gene)

Mutations that Alter DNA Sequence

Point mutations   In these type of mutations, a single nucleotide is altered or changed. These mutations are also called as micro-lesions, as they are not visible under the microscope. The mutation can be observed only when the wild type and mutant DNA sequence is compared. Point mutations do not alter the length of the DNA molecule. Point mutations are further classified into two types based upon the type of nucleotide they alter. (i.e., purine or pyrimidine) (Figure 11.1).

Transitions In this type, a purine is replaced by purine (i.e., A is replaced by G or G is replaced by A) or pyrimidine is replaced by pyrimidine (i.e., T is replaced by C or C is replaced by T).

Transversion In this type, a purine is replaced by pyrimidine or vice versa (G is replaced by C, A is replaced by C, T is replaced by A, T is replaced by G).

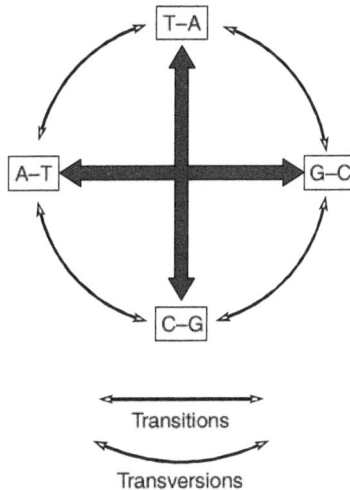

Figure 11.1 Block diagram for understanding the difference between transition and transversion

Frameshift mutations In this type of mutation, some nucleotides are added or deleted (Figure 11.2). This mutation leads to a change in number of nucleotide pairs in a gene, thus altering the reading frame of the gene. Hence they are called as frameshift mutations. These mutations are also called as intermediate lesions. Frameshift mutations are classified into two types depending upon whether they increase or decrease the gene size.

Insertion   In this mutation, one extra nucleotide is added.

Deletion   In this mutation, one nucleotide is deleted.

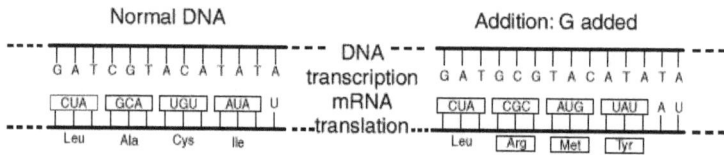

Figure 11.2   Frameshift mutation

Mutations that Alter Gene Function

Silent mutation   This is a mutation which causes a change in the sequence of DNA but does not cause any damage or loss of function of a gene. Most silent mutations occur in the third nucleotide position of a codon of a reading frame. A silent mutation has no effect on the amino acid sequence of the gene product and does not give rise to mutant phenotype. Silent mutations play an important role in RFLP studies.

Nonsense mutation   This is a point mutation that changes a codon specifying an amino acid into a termination codon. Thus a truncated gene codes for a polypeptide that has lost a segment at its carboxyl terminal. In many cases, this segment will include amino acids essential for the protein's activity and mutant phenotype results. Nonsense mutations cause premature termination of protein synthesis.

Mis-sense mutation   Mutations occurring at the first or second nucleotide position of a codon result in mis-sense mutation. This alteration will create a new amino acid instead of an old amino acid. A mis-sense mutation gives rise to a polypeptide with a single amino acid change. Whether or not this causes a mutant phenotype depends on the precise role of the mutated amino acid in the structure and/or function of the protein. Most proteins can tolerate some change in their amino acid sequence, but mutations that alter an amino acid essential for structure or function will inactivate the protein and lead to a mutant phenotype. Mis-sense mutations are further classified as conservative (or) radical depending upon the type of amino acid substituted.

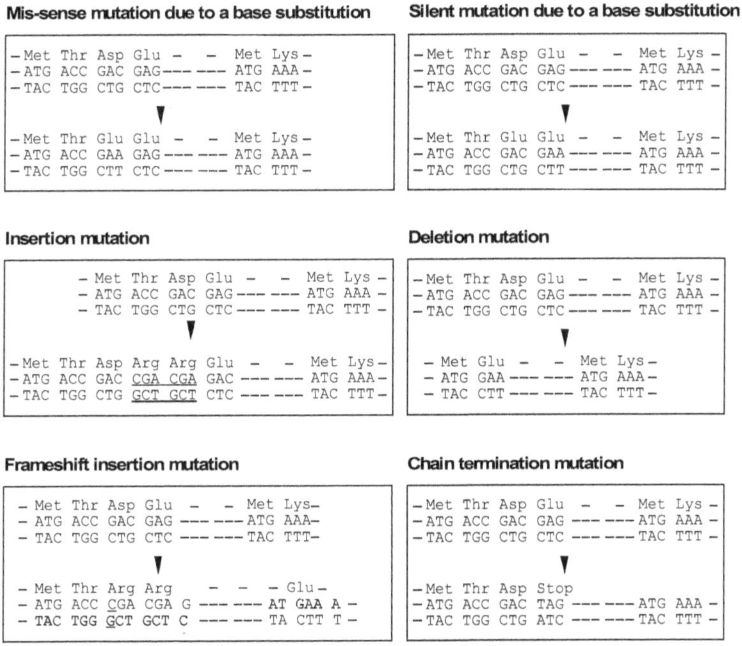

Mis-sense mutation due to a base substitution

```
-Met Thr Asp Glu -  - Met Lys-
-ATG ACC GAC GAG--- --- ATG AAA-
-TAC TGG CTG CTC--- --- TAC TTT-
              ▼
-Met Thr Glu Glu -  - Met Lys-
-ATG ACC GAA GAG--- --- ATG AAA-
-TAC TGG CTT CTC--- --- TAC TTT-
```

Silent mutation due to a base substitution

```
-Met Thr Asp Glu -  - Met Lys -
-ATG ACC GAC GAG--- --- ATG AAA -
-TAC TGG CTG CTC--- --- TAC TTT -
              ▼
-Met Thr Glu Glu -  - Met Lys -
-ATG ACC GAC GAA--- --- ATG AAA -
-TAC TGG CTG CTT--- --- TAC TTT -
```

Insertion mutation

```
  - Met Thr Asp Glu -  - Met Lys -
  - ATG ACC GAC GAG--- --- ATG AAA -
  - TAC TGG CTG CTC--- --- TAC TTT -
              ▼
-Met Thr Asp Arg Arg Glu -  - Met Lys-
-ATG ACC GAC CGA CGA GAC --- --- ATG AAA-
-TAC TGG CTG GCT GCT CTC --- --- TAC TTT-
```

Deletion mutation

```
-Met Thr Asp Glu -  - Met Lys -
-ATG ACC GAC GAG--- --- ATG AAA -
-TAC TGG CTG CTC--- --- TAC TTT -
              ▼
 - Met Glu -  - Met Lys-
 - ATG GAA--- --- ATG AAA-
 - TAC CTT--- --- TAC TTT-
```

Frameshift insertion mutation

```
 - Met Thr Asp Glu -  - Met Lys-
 - ATG ACC GAC GAG --- ---ATG AAA-
 - TAC TGG CTG CTC --- ---TAC TTT-
              ▼
 - Met Thr Arg Arg  -  -  - Glu-
 - ATG ACC CGA CGA G --- --- AT GAA A -
 - TAC TGG GCT GCT C --- --- TA CTT T -
```

Chain termination mutation

```
-Met Thr Asp Glu -  - Met Lys -
-ATG ACC GAC GAG --- ---ATG AAA -
-TAC TGG CTG CTC --- ---TAC TTT -
              ▼
-Met Thr Asp Stop
-ATG ACC GAC TAG --- ---ATG AAA -
-TAC TGG CTG ATC --- ---TAC TTT -
```

Figure 11.3 Diagrammatic representation of different types of mutations

Conservative substitutions This type of mutation causes no change in amino acid type (i.e., an acidic type amino acid, e.g. Asp is replaced by another acidic type amino acid). In all cases, the net effect on the charge of protein in the molecule is nil. Conservative mutations may or may not affect the function of protein. It depends upon the site of replacement. Certain biochemical or kinetic properties, such as V_{max}, K_m, may be affected by conservative mutations. These mutations cannot be detected electrophoretically.

Radical substitutions   In this case, amino acids having dissimilar charge (i.e., acidic group amino acid, e.g. ASP is replaced by basic amino acid, e.g. lys) or chemically reactive amino acid is replaced by non-reactive amino acid or high molecular weight amino acid is replaced by low molecular weight amino acid. These mutations result in difference in mass/charge ratio of protein hence can be detected electrophoretically. The mutation can have zero effect to full effect depending upon the type or site of replacement.

A diagrammatic representation of different types of mutations that alter DNA sequence and gene function is shown in Figure 11.3.

Mutations that Alter Gene Expression

These mutations cause difference in gene expression pattern and can be detected either by visual observations or by electrophoretic methods. Depending upon the degree of expression change, these mutations are classified into four categories.

Amorphic mutations   These mutations cause a total loss of the expression of a trait, which is due to the production of non-functional proteins or enzyme, e.g. white eye.

Hypomorphic mutations   In this type of mutation, the expression is not lost totally but is present in partial because partial enzyme activity is retained.

Hypermorphic mutations These mutations lead to an increased expression than that of wild type.

Isoallelic mutations These mutations do not affect the intensity of expression of the concerned trait. The enzyme coded by such alleles is comparable in activity to those produced by their wild alleles. The enzymes can be differentiated electrophoretically.

CHROMOSOMAL ABERRATION

Mutations which alter the chromosome structure, size or gene arrangement are chromosomal mutations. Chromosomal mutations are widely called as chromosomal aberrations. These are grouped into two broad classes based open whether they alter the structure or number of chromosomes (Figure 11.4).

Structural Chromosomal Aberrations

These aberrations cause structural abnormalities in chromosome structure. They alter the sequence or the kind of genes present in chromosome. These are further classified into four groups based upon whether they alter the gene sequences, number or location.

Deletions Deletions are caused when a chromosome breaks and loses a segment. Deletions lead to decrease in gene number. Deletion was first discovered by Bridges in 1917. Chromosomes with a deletion cannot be reverted back to wild type conditions. These are usually lethal in homozygous state. Cytologically, deletions can be detected by failure of a segment of a chromosome to pair properly. Deletions are classified into two types based upon the position where deletion occurs.

Terminal deletion It involves loss of a segment of a chromosome from the end. Normally telomere is lost, e.g. cri-du chat syndrome. This type of mutation occurs rarely.

Interstitial deletion   It involves loss of a segment between the telomere and centromere region. This type of mutation is not rare.

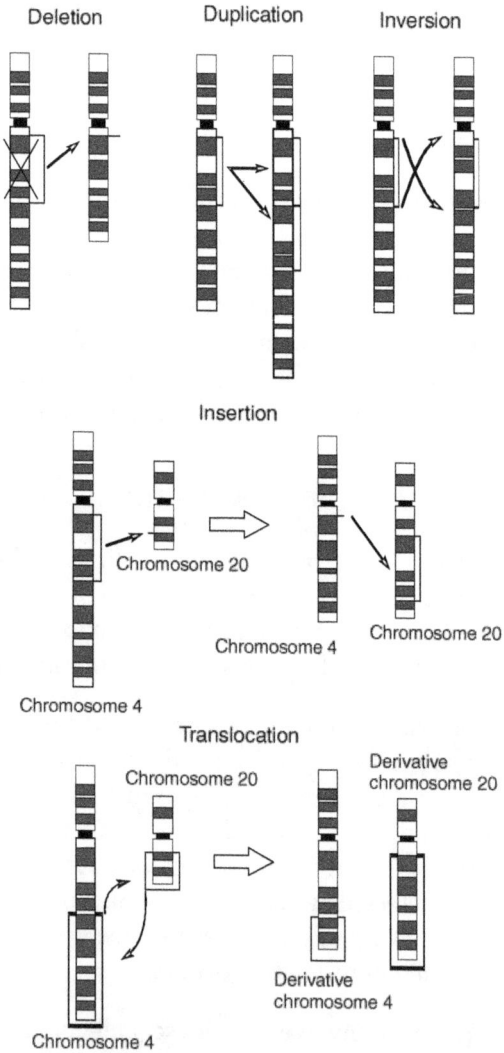

Figure 11.4   Types of chromosomal aberration

Duplication The presence of an additional chromosome segment when compared to normal chromosome is attributed to the process of duplication (Figure 11.5). Duplications were first discovered by Sturtevant and Plunkelt in 1926. Duplication means that an extra segment of gene from another or same chromosome attaches to a chromosome. In a diploid organism, if the presence of a chromosome segment in more than two copies per nucleus it is called duplication. Duplications are very important changes from evolution point of view. Duplication provides additional genetic material potentially capable of giving rise to new genes during the process of evolution. Based upon orientation and location of the new segment, duplication is further classified into four types.

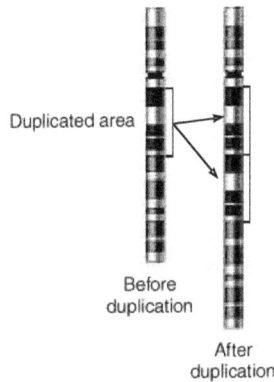

Duplicated area

Before duplication

After duplication

Figure 11.5 Duplication in the blue chromosome

Tandem duplication In this mutation the extra segment of the chromosome fragments is located immediately after the normal segment in the same orientation.

Reverse tandem duplication In this mutation, the extra segment of the chromosome fragment is located immediately after the normal segment but in the reverse order or orientation with reference to the normal similar segment.

Displaced duplication   In this mutation, the extra segment is located on the same chromosome but at a different location from the normal segment.

Translocation duplication   In this mutation, the extra segment is located on another chromosome which is unrelated to it.

Duplications do not have drastic effects like deletions. But some duplication may have distinct phenotypic effects and may be regarded as mutant allele, e.g. bar duplication in Drosophila. In duplication, the number of copies of genes will increase.

Inversion   Inversion is a phenomenon in which a portion of a gene or a group of genes which have been cut out, turn through an angle of 180° and get reattached in reverse order. Bridges first discovered inversion in 1923. In inversion the gene sequence in an inverted segment is exactly the opposite of that in its normal homologous (non-inverted) segment. In case of an inversion, there is no net loss or gain of genetic material and thus a heterozygous condition is perfectly alright, but linkage relationships change. Inversion will affect the amino acid sequence of polypeptide encoded by that gene, and thus has some genetic effect. Inversions are grouped into two groups depending upon the presence or absence of centromere within the inverted segment.

Paracentric inversion   This is an inversion in which the centromere is not involved (Figure 11.6). These only change the linkage group and sequence alignment, but do not change the chromosome structure.

Pericentric inversion   This is an inversion in which the centromere is involved (Figure 11.7). These change the chromosome structure (acrocentric to telocentric chromosome) along with modifications in the linkage group and sequence alignment. As pericentric inversions alter chromosome morphology, they play an important role in the evolution of some species, e.g. drosophila.

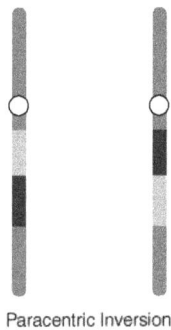

Paracentric Inversion

Figure 11.6 Paracentric inversion in the blue chromosome

Pericentric inversion

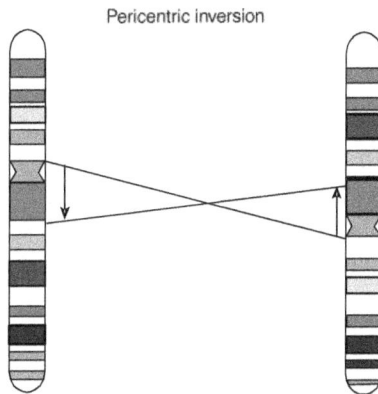

Figure 11.7 Pericentric inversion in the blue chromosome

Translocation Translocation involves transfer of a part of chromosome from one position to another chromosome. Here the original segment will not be present on the first chromosome any more (Figure 11.8). In duplication, the original segment will be present and an extra new segment will the generated, whereas in translocation, the segment is cut and joined into new position. In this case, there is no increase in the number of copies of genes. Translocation changes the linkage groups and alters the size of chromosome, e.g. Philadelphia chromosome responsible for leukemia.

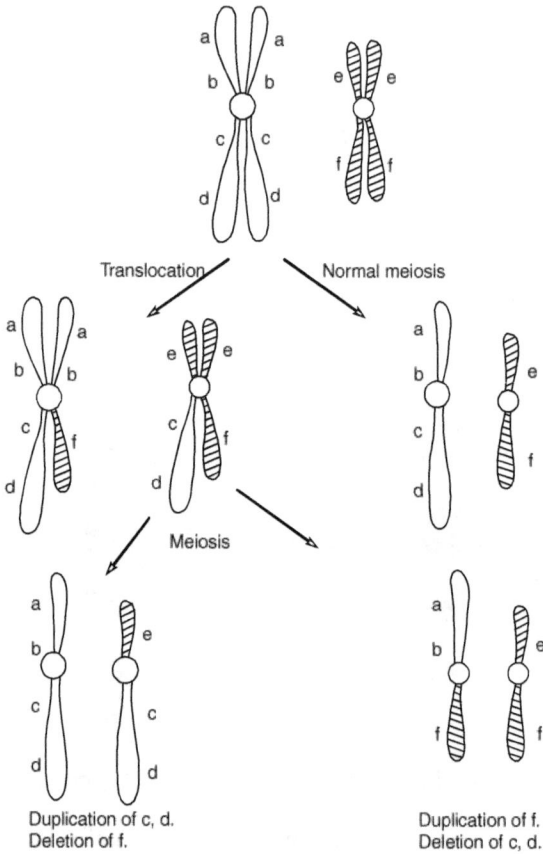

Figure 11.8   Translocation

Translocations are classified into three types.

Simple translocation   In simple translocation, a chromosome segment from the terminal region is integrated at one end of a non-homologous chromosome. Simple translocations are rare.

Intercalary translocation   In intercalary translocation, the chromosome segment is integrated within a non-homologous chromosome. These are also called as shifts.

Reciprocal translocation   In reciprocal translocation, two non-homologous chromosomes exchange segments. Each chromosome will give a segment and take a segment. Reciprocal translocation requires a single break in each of the two non-homologous chromosomes and a reunion of the chromosome segments thus produced. A non-equal reciprocal translocation between two acrocentric chromosomes may produce a dot or micro-chromosome and a metacentric chromosome. This is called as centric fusion. And a process of unequal reciprocal translocation between a mini-chromosome and metacentric chromosome is called as dissociation.

Chromosomal Aberrations Affecting the Number

In somatic cells of a diploid organism, two copies of the same genome are present, while their gametes contain a single genome. A deviation from the diploid state represents a numerical chromosomal aberration or heteroploidy and the animal or plant is called as heteroploid. Heteroploidy is grouped into two types depending upon whether the complete set of chromosomes (n) or single chromosome is involved in aberration.

Aneuploidy   A loss or gain of one or few chromosomes from the somatic chromosome number of a species is known as aneuploidy. For example, humans have 46 chromosomes of which 44 are somatic chromosomes and two are sex chromosomes. If an individual has only 45 chromosomes and he misses one somatic chromosome, then it is called as aneuploidy. Aneuploidy changes the chromosome number and does not involve the whole genome. Non-disjunction is the case of aneuploids. Aneuploidy is further classified into the following types.

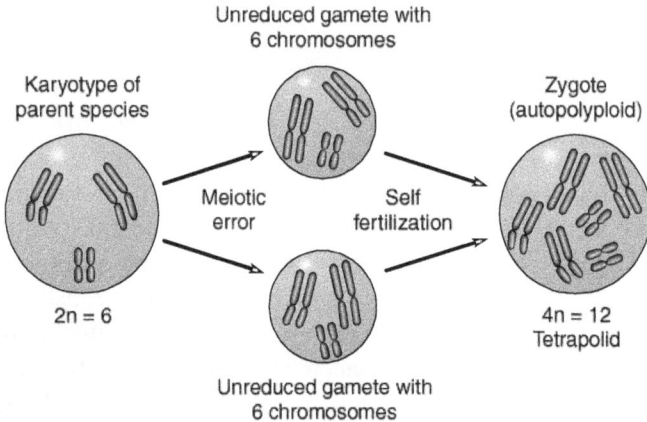

Figure 11.9   Tetraploid production

Nullisomic   In nullisomic, a pair of chromosomes is totally missing. It is denoted by (2n–2). In this the same homologous chromosome will be missing, e.g. a human cell which does not have the 13th chromosome totally, i.e., both the pairs are missing. This is always lethal in the case of humans.

Monosomic   In monosomic, a single chromosome from the pair of chromosomes is totally lost. It is denoted by (2n–1), e.g. a human cell having only one 13th chromosome without any pair.

Double monosomic   In this, two chromosomes which are non-homologous or from different pair are missing, e.g. a human cell which does not have one 13th and one 15th chromosome comes under this group. It will have one chromosome of both. It is denoted by (2n – 1 – 1).

Trisomic   In this, three copies of the same chromosome will be present. It is denoted by 2n + 1, e.g. a human cell having three numbers of the 13th chromosome (Figure 11.10).

Tetrasomic   In this one extra chromosome pair will be present and is denoted by 2n + 2, e.g. a human cell having four numbers of the 13th chromosome.

Double transonic   In this, two extra chromosomes, which are non-homologous or are a different pair, will be present. It is denoted by 2n + 1 + 1.

These mutations lead to production of plants with reduced vigour and fertility. Plants with aneuploidy mutations are used in studies related to chromosome substitution and to locate the effect of individual chromosomes in the cell.

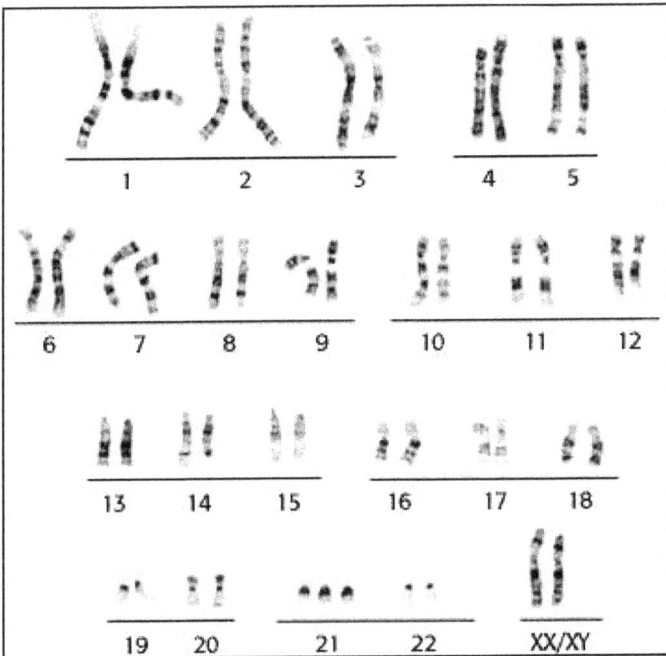

Figure 11.10   Human female karyotype (47 {+21})

Euploidy   Euploids have one or more number of complete genome (chromosome set). In the diploid conditions, the genomes are designated as 2X or 2n. Euploids are also designated with reference to diploid stage (3X, 4X, 3n or 4n). Euploidy is divided into two groups based upon the number of chromosomes or genomes present.

Haploids   These denote the presence of single copy of a genome. These represent the gametic chromosome number of a species. Haploids are represented by "n". These mutants are highly sterile and used in production of the homozygous lines. In haploids, none of the chromosomes has two homologues. Hence, all chromosomes are unpaired during meiosis and are present as n equivalents.

Polyploidy   Presence of more than two genomes (i.e., more than 2n number) is called as polyploidy. Polyploidy is further divided into two groups based upon whether they contain the same genome or different genomes.

1. *Autopolyploidy*   Autopolyploids have three or more than three genomes of same type. For example, if a chromosome of a diploid cell of onion plant is fused with haploid or diploid cell of onion plant. Then they are called as autopolyploids. Autopolyploids normally have genome from the same species. Autopolyploids are further classified into further groups depending upon the number of genomes present such as triploid (three pairs), tetraploids (four pairs) (Figure 10.9), pentaploids (five pairs), etc.

2. *Allopolyploidy*   Allopolyploids contain two or more distinct or different genomes. These are formed from sets of chromosomes of different species. These are also called as amphidiploids. Allopolyploid seems to play a vital role in speciation of plants. Artificial production of allopolyploids is called as the synthesis of somatic hybridization and the allopolyploids are called as synthetic allopolyploids. Allopolyploids are more vigorous than parental species and have unpredictable combinations of parental features.

ORGANISM LEVEL MUTATIONS

Most of the mutations do not produce any phenotypically differentiating abnormalities, but some mutations can cause changes in the phenotype or ability to resist a substance (toxin) or inability to code for some protein. Such mutations are called as phenotypic mutations. Phenotypic mutations can be observed by the naked eye. Phenotype mutations are further classified into various types based upon the ability to survive in the presence of mutation.

Lethal mutants In lethal mutation, the organism dies due to mutation. The mutation abolishes the gene function totally. Many lethal mutations affect some vital functions of the organism. Depending upon the degree of lethal effect, lethal mutations are classified as sublethal, super-vital and conditional lethals.

Super-vital mutations These mutations enhance the survival of the individual carrying them in the appropriate environment or stress. Antibiotic resistance and stress tolerance comes in this category.

Conditional lethals The lethal mutations which require specific conditions for expressing lethality are called as conditional lethals. These mutations are further classified into various types based upon the factor which suppresses or enhances the lethality.

Auxotrophic mutants This type of mutation is involved in the synthesis of the essential metabolite. These mutants can be maintained alive by giving the essential metabolite extra in the food or media. Some mutants which do not require a specific metabolite when compared to wild type are called as prototrophs. These are also called as biochemical mutants.

Temperature-sensitive mutants This type of mutants survive at one temperature and die when exposed to another temperature. Most temperature-sensitive mutants are sensitive to heat. These mutants arise due to mutations in heat shock proteins or chaperone genes.

A mutation can occur in nuclear genes (Nuclear mutation) or in chloroplasts, mitochondria or cytoplasmic genes (cytoplasmic mutations). Nuclear mutations can be autosomal mutations (occur in autosomes) or sex-linked mutations (occur in allosomes) or holandric mutations (occur in sex chromosomes present in one sex individual, e.g. Y-chromosome). Mutations which occur in somatic cells and are not transferred to the progeny are called as somatic mutations. Those mutations which occur in germinal cells and which are transferred to the progeny are called as germinal mutations.

MOLECULAR BASIS OF MUTATION

Even though DNA replication is based upon specific base pairing theoretically, in practice there are possibilities of occurrence of errors. Due to the non-static nature of the bases present in DNA, the hydrogen atoms present in the amino group can move from one position in a purine or pyrimidine to another position. Such chemical fluctuations are called as tautomeric shifts. Tautomers are structural isomers that exist in equilibrium with one another. Tautomers of a single compound have the same chemical formulae but slightly different molecular structures. In pyrimidines (T and C), the hydrogen atoms shift from 3 atom (N) to 4 atom (0), whereas, purines (A,C) in the shift occurs between the number 1 and number 6.

Although tautomeric shifts are rare, the more stable keto form (thymine and guanine) and amino forms (adenine and cytosine) may undergo tautomeric shifts to a less stable enol and imino form for only very short period of time. But presence of such bases during DNA replication or repair, might result in a mutation. When the bases are present in their rare imino or enol states, they can form adenine–cytosine and guanine–thymine base pairs.

Mutations resulting from tautomeric shifts in the bases of DNA involve the replacement of a purine in one strand of DNA with the other purine and the replacement of a pyrimidine in the complementary stand with the other pyrimidine. Such base pair substitutions are called as transitions. Proton migrations within base pairs can set a chain of modification that end up in mutation. Methylated guanine undergoes tautomeric shifts and pairs with thymine and leads to GC–TA transversions. Transversions occur by apposition of two errors—proton migration in one and 180° rotation around glycoside bond (syn). Tautomeric synadenine (A*) may pair with Syn A which leads to AT–TA transversions. Similarly tautomeric neoguanine (G*) may also pair with syn G which leads to GC–CG transversion. Even though all types of mutations are seen in various organisms, the probability of base deletions or addition is more than that of other mutations. Till now there is no clear explanation why this occurs. One point is clear, we have to learn a great deal about the cause and mechanism of mutation.

Mutation frequency Mutation frequency or rate is defined as a probability of mutation per cell per generation, independent of exact physiological conditions and stage in the life cycle. Each gene probably has its own characteristic mutation behaviour. Some genes undergo mutations more frequently than others in same organism. The mutation rate in bacteria per gene is the order of 1 mutation per 10^{-7} generation per cell generation. Mutation frequency of various types of spontaneous mutations depends upon the accuracy of the DNA replication machinery, efficiency of the numerous mechanisms to repair damaged DNA and the degree of exposure to various mutagenic agents present in the environment.

MUTAGENIC AGENTS OR MUTAGENS

An agent that has the ability to produce mutation is called the mutagen. A mutagen can be a physical agent (e.g. UV, X-rays) or a chemical. Physical mutagens are radiations of fixed specific wavelength and energy. Some radiations such as X-rays and γ-rays, are capable of causing ionization (release radicals or ions) in matter through which they pass. Radiations are of two types. 1) Non-particulate, e.g. X-rays. 2) Particulate radical, e.g. α, β, γ radiation.

UV Light

Ultraviolet rays are electromagnetic and non-ionizing radiations. Because of their lower energy, they penetrate the surface tissues of multicellular animals and plants. UV is readily absorbed by certain substances like purines and pyrimidines, which then enter into a more reactive or excited state. The two major products formed by absorption of UV by pyrimidine are pyrimidine hydrates and thymine dimers. UV acts on any 'C' atom of pyrimidine of DNA in water solution and adds a water molecule across the double bond. But thymine dimerization is probably the major mutagenic effect by UV. UV light produces thymine dimer mainly by linking adjacent thymine bases in the same strands of DNA via C–C bonds. These thymine dimers block the transcription and replication. Normally thymine dimers are repaired by DNA repair mechanism, and only a small fraction of the dimer results in mutations. The relationship between mutation rate and UV dosage is highly variable depending on the type of mutation, the organism and conditions employed for the study. Electromagnetic radiation with a wavelength of 0.05–10 Å are emitted in quanta. X-rays affect the chromosomes by directly breaking them or altering the DNA base pair. Chromosomes are extremely sensitive to breakage during meiotic prophase.

Chemical Mutagens

Chemical mutagens can be divided into two classes—those mutagens which can cause mutations to both replicating and non-replicating DNA, are called as class I mutagens, those chemicals, which affect replicating nucleic acids are called as class II mutagens. The molecules of class II mutagens look like nucleic acid; hence they are incorporated into the replicating DNA molecule. The most prominent mutagenic members of this class are base analogs and dyes.

Base analogs Base analogs are mutagenic and have a structure sufficiently similar to the normal bases so that they are metabolized and incorporated into DNA during replication, but sufficiently different such that they increase the frequency of mis-pairing and thus cause mutation. The two most commonly used base analogs are 5´-bromouracil and 2´-aminopurine. 5´-bromouracil (5-BU) is a thymine analog. CH_3 in 5´ position of thymine and bromine at 5´-position in 5-Bu have similar effect. But probability of tautomeric shifts increase because of bromine molecule. After a tautomeric shift to its enol form, 5´-bromouracil pairs with guanine and thus causes A.T to G.C transitions (Figure 11.11). 2-aminopurine can pair in its normal tautomeric form with two bases, with thymine by two hydrogen bonds and with cytosine by one hydrogen bond. The normal base that pairs with base analogues, 2-aminopurine and 2, 6-diaminopurine is not known.

Base analogs cause transitions in the bidirectional mode. Hence mutations induced by these substances can be reversed by using the same compounds, e.g. 5´-bromouracil, 5-chlorouracil, 5-iodouracil, 2-aminopurine, 2, 6-diaminopurine, etc.

Acridine dyes Acridine dyes bind directly to the DNA by using their positive charge. Positively charged acridines intercalate or sandwich themselves between the stacked base pairs in DNA. Acridine dyes separate two bases by 6.8 Å, thus a base will be missed. Acridine dyes can insert or delete only one base pair in

DNA, and thus result in frameshift mutations. The examples of this class of mutagens are nitrous acid (HNO_2), hydroxylamine, hydrazine, H_2O_2, etc.

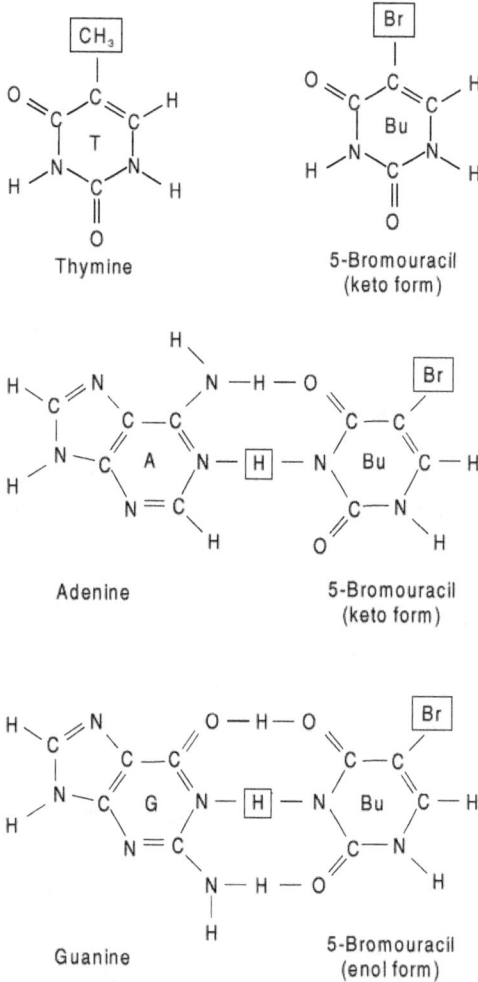

Figure 11.11   Molecular mechanism of base pairing between base and mutagens

Nitrous acid   Nitrous acid is a very potent mutagen that acts directly on either replicating or non-replicating DNA by oxidation or deamination of the bases that contain amino groups (adenine, guanine and cytosine). Conversion of the amino groups to keto groups changes the hydrogen-bonding potential of the bases. Adenine is deaminated to hypoxanthine, which then pairs preferentially with cytosine in the place of thymine. Cytosine is deaminated to uracil, which now pairs with adenine in the place of guanine. Deamination of guanine has zero effect, as deaminated guanine also pairs with cytosine. Since the deamination of adenine leads to AT. GC transition and the deamination of cytosine results in GC. AT transitions, nitrous acid induces transitions in both directions. Nitrous acid also causes inter-strand cross-linking of DNA. The DNA strands fail to separate and there is no DNA duplication, which is lethal or deleterious.

Hydroxylamine   Hydroxylamine (NH_2OH) reacts with pyrimidine bases. Its effect is strong on cytosine while on uracil the effect is less. Hydroxylamine breaks and removes pyrimidine ring of uracil thus producing phosphoribosyl urea and 5´-isoxasolone with cytosine, hydroxylamine finally produces hydroxyl amino (–NOH) derivative, which might be responsible for base pair change (GC–AT pair). Mutations induced by hydroxylamine cannot be reversed with hydroxylamine.

Alkylating agent   Alkylating agents carry one, two or more alkyl groups in reactive form, which are capable of being transferred to other molecules where electron density is high. One major mechanism of mutagenesis by alkylating agents involves the transfer of methyl or ethyl group to the bases such that their base-pairing potentials are altered and transitions result (Figure 11.12). Dimethyl sulphate upon reaction with guanine in DNA produces 7-methyl guanine, which may pair with thymine. The result is transition of GC to AT. Alkylating agents may cause cross-linking of DNA and thus cause lethal effect by inhibiting DNA duplication because of cross-linking. Alkylating agents may readily react with PO_4 group and may break the sugar-phosphate backbone of DNA. This may induce large alterations. Difunctional alkylating agents cause cross-linking of DNA, induce chromosome breaks and chromosomal aberrations. Alkylating agents as a class exhibit less specificity in their mutagenic effect than base analogs or dyes.

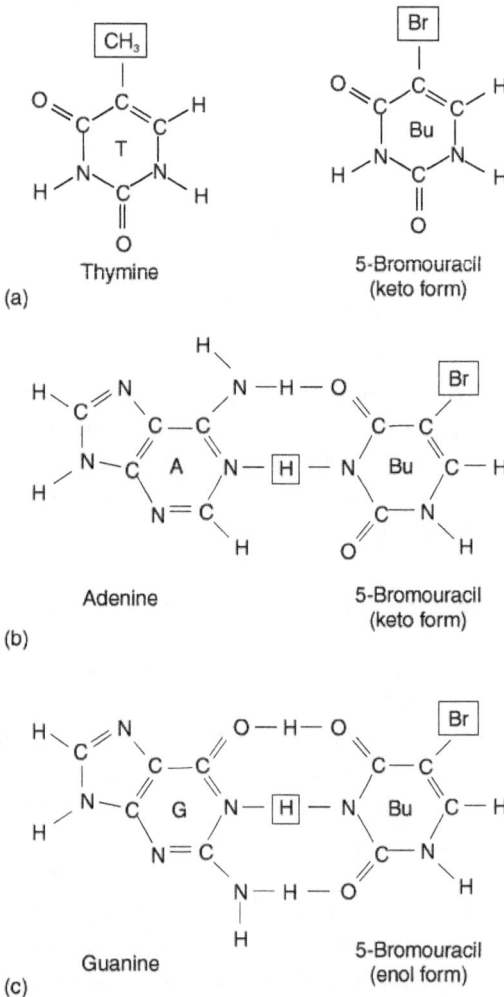

Figure 11.12   Interaction of alkylating agent with mutagen

They induce all types of mutations, (transitions, transversions, frameshifts and even chromosome aberrations) with various relative frequencies depending on the specific alkylating agent.

Antimutagens

Antimutagens are compounds which are nontoxic, non-mutagenic and suppress mutagenicity of a chemical. Antimutagens show positive effect on spontaneous mutations but not on chemically induced mutations. Interestingly, a single antimutagen shows positive effect in one organism and negative in other. Antimutagens are classified into two categories.

Desmutagens

These are compounds which directly inactivate the mutagens, e.g. vitamins A, C, and E, chlorophyllin (sodium or copper salts of chlorophyll).

Bio-Antimutagens

These are compounds which act on suppressors of cellular mutagenesis, e.g. 5-fluorouracil, fluorodeoxyuridine. Bio-antimutagens may act in various ways, i.e., prevent error-prone SOS repair or SOS repair with increased fidelity.

DETECTION OF MUTATIONS

For many years it has been recognized that most of the compounds and physical light sources have strong mutagens. Hence most governments have made it mandatory that before a molecule or source is released out, it has to be certified whether it can cause mutations or not. Various tests are available to assess the mutagenicity of a mutagen against various organisms.

Replica Plating Method

In this method, bacteria are grown for a fixed amount of time along with a mutagen. Then the bacteria are plated on nutrient media supplemented with vitamins, amino acids, etc. This plate is used as master plate. Replicas are made by pressing velvet-covered wooden blocks against the master plate and then pressed in the same orientation onto the surface of petri plates containing fresh nutrient media supplemented with vitamins and amino acid, expect one vitamin or amino acid. Then the numbers of mutant colonies or auxotrophs generated are counted. In some cases a known auxotroph is taken and checked for reversion to wild type after mutagen treatment.

Ames Test

This method is used to detect the mutagenic effect of various commercial and pharmaceutical products. Bruce Ames and his associates have developed this sensitive technique, which is rapid and inexpensive and is called as Ames test. Ames procedure involves the use of specially constructed strains of *Salmonella typhimurium*, which requires histidine amino acid. Ames test relies on conversion of auxotrophic mutants (his-mutants) back to wild type or prototrophs. The reversion of these auxotrophic mutants to prototrophs either spontaneously or induced by various agents can easily be monitored by placing a known number of mutant cells in a petri plate containing medium supplemented with a trace of growth factor required by the auxotroph (enough to support a few cell divisions but not enough to allow the formation of visible colonies). After growth, the number of colonies of prototroph bacteria formed is counted. The frequency of reversion induced by a particular substance can be compared directly with the spontaneous reversion frequency to obtain an estimate of its mutagenicity.

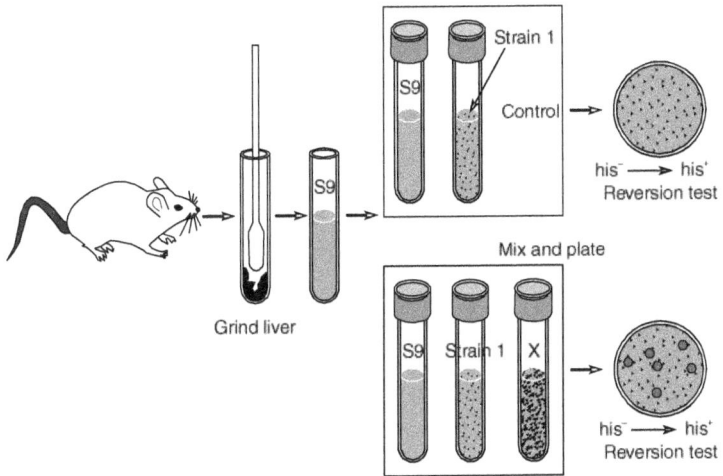

Figure 11.13   Ames test procedure

To improve the efficiency of the test, bacterial strains defective in DNA repair and enhanced permeability are used. In addition, the substances to be tested are pre-incubated with liver extracts in order to stimulate mammalian metabolic activity and to discover whether such substances as metabolized in the body produce mutagenic activity. The ability of a mutagen to induce different types of mutations can be assessed by using a set of test strains that carry different types of mutations. Strain TA 100 with a transition is highly sensitive to reversion by base pair substitutions, whereas strain TA 1535 and TA 1538 are sensitive to reversions by frameshift mutations.

CIB Technique

This method was invented by Muller. In this method, females containing one normal X-chromosome and another X-chromosome (CIB) containing extra 3 genes are used for the analysis. Out of the 3 extra genes, one gene suppresses crossover (c), the other is a recessive lethal (L) in heterozygous condition, and the last gene is semi-dominant marker, Bar (B) gene. Females containing CIB chromosome are called as CIB stock drosophila. The normal males are exposed to mutagenic source for a fixed period and then mated to the CIB stock drosophila. Males containing CIB chromosome will die due to the effect of lethal genes, whereas normal males and females both normal and with CIB will survive. Females with CIB chromosomes and identified by barred phenotype are selected and crossed to normal males. In this next generation 50% of males (which have received the CIB gene) will die. If mutation has occurred in normal X-chromosome then even the normal male (without CIB gene) will die. If no mutation has occurred all the other 50% of males will survive. The frequency of lethal mutations can be accurately scored in large samples.

APPLICATIONS OF MUTATIONS

Mutations are invaluable to the process of evolution since they provide the raw material required for its occurrence. Even though most mutations make the organism less efficient and are thus disadvantageous, the possibility of developing new traits through induced mutations has intrigued many plant breeders. Many plant strains were improved for higher yield, higher resistance against disease and pest by using mutations.

One application of induced mutations comes from the concentrated efforts to improve the yield of penicillin by the mold penicillium. When penicillin was first discovered, the yield was low and production was seriously limited. Then millions of spores were irradiated and selected for higher yields. Such mutants which produce penicillin more than the average have proven invaluable in the commercial production of this important antibiotic.

Mutations have been used extensively to elucidate the pathways by which biological processes occur. Metabolisms occur via sequences of enzyme-catalysed reactions. By isolating and studying mutations in the genes coding for the enzyme involved, the sequence of steps in a pathway can often be determined.

Morphogenesis frequently involves the sequential addition of proteins in the formation of specific 3-D structures. Again, the sequence of protein additions can often be determined by isolating and studying mutant organisms with mutations in the genes coding for the proteins involved.

REVIEW QUESTIONS

1. Describe in detail about DNA molecule.

2. Write in detail about chromosomal mutation.

3. Explain phenotypic mutation in detail.

4. Describe the mutation that alters gene function.

5. Write the difference between DNA and chromosomal mutation.

6. Write the difference between euploidy and aneuploidy.

7. Explain the difference between transition and transversion.

8. Describe in detail about AMES test used to detect mutagenicity.

9. Explain the molecular basis of mutation.

10. Write short notes on the following.

i.	Transitions	viii.	Conservative substitutions
ii.	Antimutagens	ix.	Radical substitutions
iii.	Transversion	x.	Inversion
iv.	Alkylating agent	xi.	Deletions
v.	Silent mutations	xii.	Duplications
vi.	Acridine dyes	xiii.	Translocation
vi.	Nonsense mutations	xiv.	CIB technique
vii.	Mis-sense mutations		

12

BACTERIAL RECOMBINATION

Bacteria divide by simple fission, and are usually haploid with multiple copies and types of genetic constituents, e.g. chromosomes, episomes, plasmids, etc. Bacteria do not undergo mitotic or meiotic cycles; hence they do not undergo sexual reproduction process involving gametic union. The genetic recombination events associated with sexual reproduction like segregation, chromosomal exchange, etc. are not integral parts of the life cycle of bacteria. Recombination, however is undoubtedly important in the evolution of eukaryotes, so is it with bacteria. Although they do not undergo sexual reproduction by the fusion of haploid gametes, bacteria and viruses do undergo processes that incorporate genetic material from one cell or virus into another cell or virus forming recombinants. Actually, bacteria have three different methods to gain access to foreign genetic material; transformation, conjugation and transduction. The most obvious difference between these three processes is the mode of transfer of DNA from one cell to another. Transformation involves the uptake of naked DNA molecules from one bacterium to another. Transduction occurs when bacterial genes are carried from one bacterium to another by a bacteriophage. Conjugation is the process during which DNA from a donor or male cell is transferred to a recipient or female cell through a specialized sex pilus or conjugation tube.

The three modes of recombination in bacteria can be distinguished by two simple criteria—sensitivity to the presence of deoxyribonuclease (DNase) and dependence on cell contact. Recombination by conjugation is not sensitive to DNase and requires the contact of donor and recipient cell whereas transformation is sensitive to DNase treatment and requires no contact of donor and recipient cell. Transduction is very peculiar in that it does not require any contact of donor or recipient cell and is not sensitive to DNase treatment.

CONJUGATION

Conjugation is a type of recombination in bacteria. During conjugation, DNA is transferred from a donor cell (called as male cell ♂) to the recipient cell (called as female cell ♀) through a specialized intracellular connection or conjugation tube that forms between them. The transfer of genetic information is one-way transfer (during conjugation) rather than a reciprocal exchange of genetic material. Cells that have the capacity to serve as donors during conjugation are different due to the presence of specialized cell surface appendages called F pili. F pili are 2–3 mm long, ~8 nm diameter with a 2 nm axial hole. A typical F^+ cell has 2–3 pili. Genes present on a plasmid called as F^+ plasmid or F-factor control F-pili synthesis. This F^+ plasmid is 94.5 kb is length with more than 60 known coding genes. This F^+ plasmid can exist at a level of one copy per bacteria in free form and replicates as F^+ cell. But sometimes F^+ plasmid gets integrated into the bacterial chromosome, ceases to exist in free form and replicates only when bacterial chromosome replicates. An interesting thing is, when F^+ plasmid integrates into the host chromosome, it loses its ability to transfer F+ plasmid genes and instead transfers the host chromosomal genes. Such mutant F+ strains are called as Hfr or high frequency recombinant strain or cell. The F+ can integrate into the host chromosome at any one of many sites. The integration of the F factor is believed to be mediated by short DNA sequences called IS elements.

Molecular Mechanism of Conjugation

The mechanism of DNA transfer during conjugation is complex and well coordinated. Proteins traY and traI nick one strand of the F+ factor or Hfr chromosome at a specific locus called origin of transfer (*oriT*). The 5′-end of the nicked strand progressively separates from its complementary strand and is transferred into the F− cell through the conjugation tube (Figure 12.1). traY/tra I, multimer protein binds to *oriT* after nicking and facilities the separation of the DNA molecule. The separation of the single-stranded DNA from the complementary strand is thought to be coupled with rolling circle replication model. Surprisingly, only one unit length of F factor is transferred from an F+ to F− cell even when it is linked with rolling circle replication, which can produce concatemers or multiple copies of the DNA molecule. The mechanism is clearly not understood. About 1200 bp per second are separated and is independent of DNA replication. The separated single strand is transferred into the F− cell with the 5′ end first. The single strand transferred into the F− cell serves as a template for the synthesis of its complementary strand.

Conjugation is controlled by a 33 kb region of the F plasmid called transfer region (tr). This region contains ~40 genes, which are named as *tra* and *trb* loci. These genes are arranged in 3 transcriptional units (*traJ, traM* and *tra Y*-I) (Figure 12.2). *traJ* encodes regulator or repressor protein, which control the expression of *traM* and *tra Y*-I. Gene *finP* codes a regulator anti-sense RNA that turns off *traJ*. Interestingly gene *finP* is under the control of another gene *finO*. Gene *traA* encodes the monomeric protein pilin, which is assembled into F-pili. The assembly of F pili is governed by at least 12 *tra* genes. Conjugation begins when the tip of F-pilus of a cell comes in contact with F⁻ cell. Interaction of F-pili with F+/Hfr cells is prevented by two genes, *traS* and *traT*, which encode for surface exclusion proteins that make the cell a poor recipient for such contacts. Once the mating is initiated, pili disassemble, the conjugating cells come

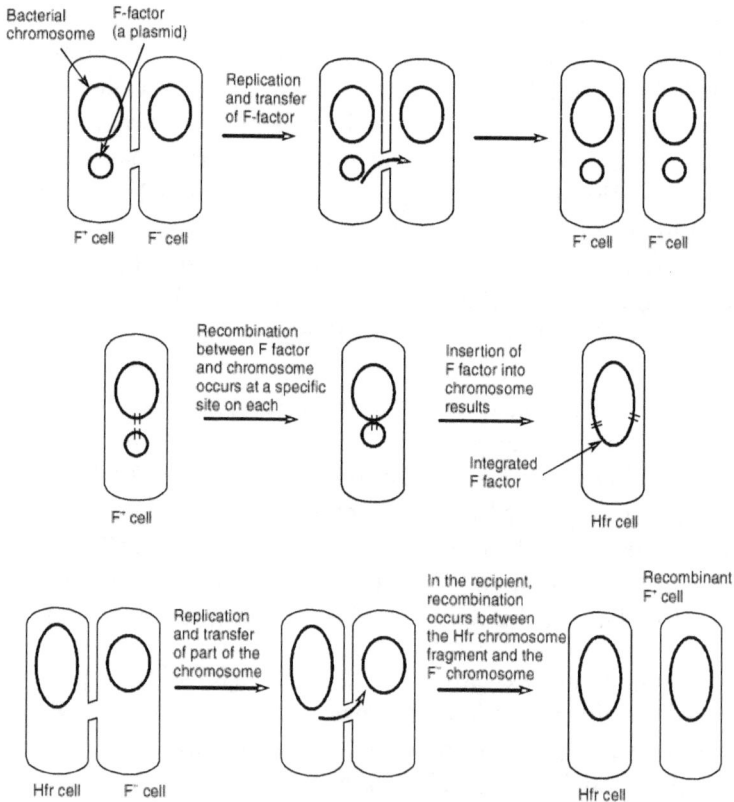

Figure 12.1   Process of production of F⁺ and Hfr strains

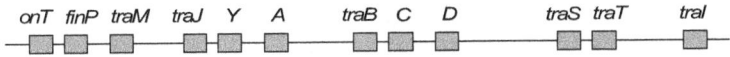

| onT | finP | traM | traJ | Y | A | traB | C | D | traS | traT | traI |

Figure 12.2 Gene arrangement in F plasmid

closer and a conjugation bridge is formed by *traD* gene products. Protein *traM* recognizes the formation of mating pair, following which *traY* binds near *oriT*. This causes *traI* to bind and nic *oriT* at a unique site called *nic*. *traI* is a relaxase (helicase activity). The different gene products and their role in conjugation are listed in Table 12.1.

Table 12.1 Gene products and their role in conjugation

Gene product	Function
oriT	Origin of transfer
traP	Turns off/on *traJ* by coding regulator protein
traM	Codes for protein involved in recognition of F strain
traJ	Encodes a regulator which turns on *traM* and *traY*-I
traY	Nicks the DNAL of F-factor at *oriT* site
traA	Encodes the monomeric protein pilin
traD	Formation of conjugation tube
traS	Codes for surface exclusion proteins F+/Hfr
traT	Codes for surface exclusion proteins in F+/Hfr

Mapping of Chromosome by Conjugation

Elie Wollmann and Francois Jacob developed the technique of chromosome mapping by conjugation. In this method, two auxotrophs are conjugated for a defined length of time and then their conjugation was interrupted by placing them in a waring blender. Then it was checked whether a given gene had been transferred.

Consider an F⁻ leu⁻ lac⁻ gal⁻ str auxotroph (an F⁻ auxotroph incapable of growing on media lacking the amino acid leucine, lactose or galactose, but resistant to streptomycin). Imagine that we conjugate it with an Hfr lev⁺ lac⁺ gal⁺ str (streptomycin-sensitive) auxotroph for two minutes. Then we select str⁺ leu⁺ recombinants by growing them in medium containing streptomycin and lacking leucine. But after only two minutes of conjugation, no str recombinants that are leu⁺ are formed. This tells us that in two minutes of the transfer no *leu* genes have been transferred from the streptomycin-sensitive Hfr cell to the streptomycin-resistant F⁻ cell. On the other hand, if we allow conjugation for four minutes, a significant number of this recombinant will arise. Therefore, a transfer of the *leu* genes takes about four minutes after conjugation begins. Taking the procedure a step further, if conjugation proceeds for ten minutes, some leu⁺ lac⁺ str recombinants appear. This means that the Hfr cells begin to transfer its *lac* genes about ten minutes after the onset of conjugation. Furthermore, if conjugation proceeds for twenty minutes, some leu⁺ lac⁺ gal⁻ str recombinants appear. Therefore the Hfr cell has begun to transfer its *gal* genes by this time.

Since we are following the transfer of genes from the Hfr donor to the F⁻ recipient, it is important to select only recipient cells after conjugation. Streptomycin sensitivity of the Hfr cells allows us to kill them with streptomycin. Not all Hfr strains transfer the *leu* genes first, each strain has its own characteristic starting point. Moreover, some strains transfer genes in an order opposite to what has just been described. Thus, *gal* would be transferred before *lac* and *leu* would be transferred last of all. Even though the order varies, all the genes are in order (Figure 12.2). For example, *leu* is transferred first and *azi* gene is transferred last by Hfr 1, whereas these two genes are transferred one after the other (*azi* first then *leu*) in Hfr2, Hfr3, Hfr4 (Figure 12.3). All we must do to make these two orders identical is to form a circle. Then the only difference will be location of origin of replication, which determines which gene will be transferred first. (Now look at the order of Hfr5, 6 and 7, you will see that the order of transfer of these two genes is reversed *leu*, then *azi*. Again, the order of gene transfer by these two strains can be reconciled by forming a circle of each and twining one of them over as you would flip a coin.)

Hfr 1   –Ori-Leu-Thr-Thi-Met-Ile-Mtl-Xyl-Mal-Str-Gly-His-Trp-Gal-Pur-Lac-Pro-Ton-Azi

Hfr 2   –Ori-Pro-Ton-Azi-Leu-Thr-Thi-Met-Ile-Mtl-Xyl-Mal-Str-Gly-His-Trp-Gal-Pur-Lac

Hfr 3   –Ori-Pur-Lac-Pro-Ton-Azi-Leu-Thr-Thi-Met-Ile-Mtl-Xyl-Mal-Str-Gly-His-Trp-Gal

Hfr 4   –Ori-Met-Thi-Thr-Leu-Azi-Ton-Pro-Lac-Pur-Gal-Trp-His-Gly-Str-Mal-Xyl-Mtl-Ile

Hfr 5   –Ori-Ile-Met-Thi-Thr-Leu-Azi-Ton-Pro-Lac-Pur-Gal-Trp-His-Gly-Str-Mal-Xyl-Mtl

Hfr 6   –Ori-Ile-Met-Thi-Thr-Leu-Azi-Ton-Pro-Lac-Pur-Gal-Trp-His-Gly-Str-Mal-Xyl-Mtl

Hfr 7   –Ori-Ton-Azi-Leu-Thr-Thi-Met-Ile-Mtl-Xyl-Mal-Str-Gly-His-Trp-Gal-Pur-Lac-Pro

Figure 12.3   Arrangement of genes in different Hfr strains

SEXDUCTION

Sometimes the inserted F-factor is occasionally excised from Hfr chromosomes giving rise to F⁺ cells. But sometimes this excision event produces an F = factor that carries host genes. Such F-factors are called as F′ (F-prime) plasmids. F′ plasmids were first discovered by E.A. Adelberg and S.N. Burns in 1959. The F′ plasmid may carry an insert or chromosomal gene ranging in size from a single gene to about ½ of the bacterial chromosomes. The transfer of F′ plasmid into F strain is believed to be similar to that of transfer of F⁺ plasmid into F⁻ strain. The F-factor has some interesting and useful properties. Alternatively, recombination between the F′ factor and the bacterial chromosome may lead to the integration of the F factor into the latter; such a gene transfer is called sexduction or F–duction. Thus sexduction produces a partial diploid for the segment of bacterial chromosome present in the F′ factor. Because of the partial diploidy resulting from sexduction, it provides an important method for determining dominance relationships between alleles and defining genes by complement test.

TRANSDUCTION

Transduction is a process in which a bacteriophage particle carries a segment of the chromosome from one bacteria (the donor) to another bacteria (the recipient). Transduction was discovered by N.Zinder and J. Lederberg. Transduction occurs by two methods, namely, generalized transduction and specialized transduction.

Transduction process in which the viral DNA is inserted into the bacterial chromosome at a random site is known as generalized transduction (Figure 12.4). Thus a random or nearly random segment of bacterial DNA is packed up during phage maturation in place of or along with the phage chromosome and transferred into a new bacteria. Phages that carry such a chromosome are called as transducing particles. Transducing particles of phages are classified into two types based upon the interactions with the bacterial cell. Virulent phages always multiply and lyse the host cell after infection. Temperate phages have a choice between two lifestyles after infection. They can either enter lytic or lysogenic pathway. Not all virulent phages mediate transduction. Bacteriophage (T3, T4 and T6) degrades the host DNA and reutilizes the mononucleotides produced, for the synthesis of phage DNA whereas other bacteriophages (T2, T5 and T6) do not degrade the host DNA at all. In such a case, a fragment of DNA from host can be packed, since the host chromosome is too large to be packaged intact. After a transducing phage injects a fragment of host DNA into a recipient cell, that DNA may be integrated into the host chromosome or remain free in the cytoplasm. Cells that carry non-integrated transducing fragments are called abortive transductants. They are partially diploid and can be used to carry out complementation tests. As the gene is not integrated, it will be transmitted to one progeny cell during each cell division.

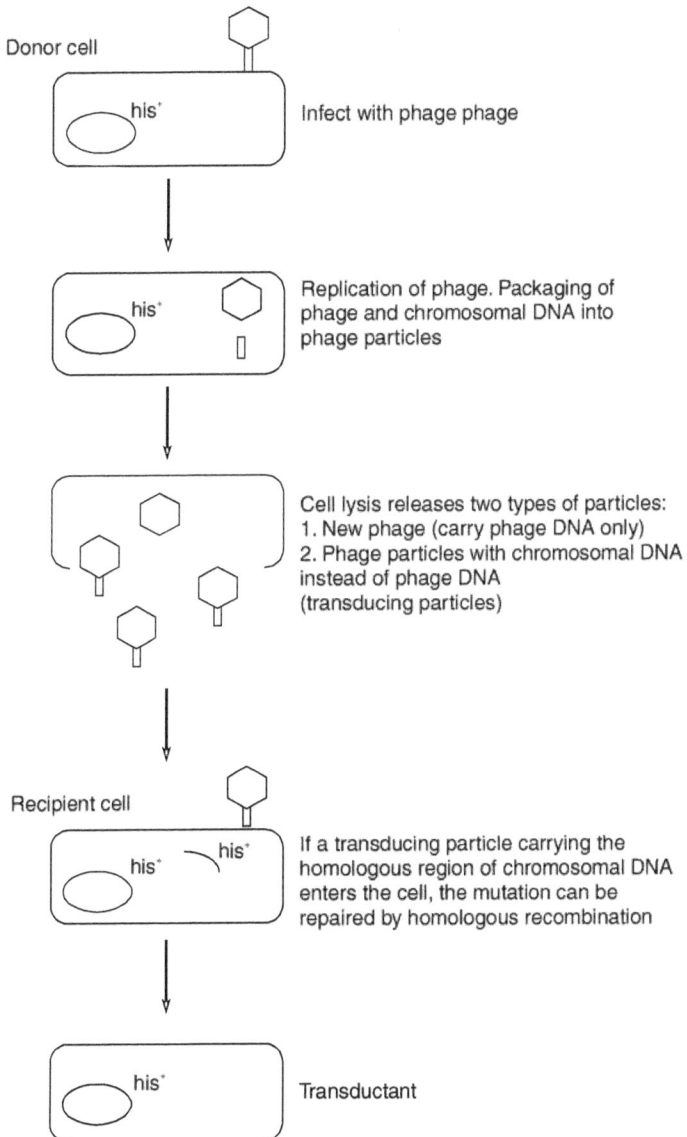

Donor cell

his˙ Infect with phage phage

his˙ Replication of phage. Packaging of phage and chromosomal DNA into phage particles

Cell lysis releases two types of particles:
1. New phage (carry phage DNA only)
2. Phage particles with chromosomal DNA instead of phage DNA (transducing particles)

Recipient cell

his˙   his˙ If a transducing particle carrying the homologous region of chromosomal DNA enters the cell, the mutation can be repaired by homologous recombination

his˙ Transductant

Figure 12.4   Generalized transduction

Specialized Transduction

Specialized transduction is mediated by temperate bacteriophages whose chromosomes are able to integrate at one or a few specified attachment sites on the host chromosome. Like episomes the phage DNA can replicate independently or along with host chromosome replication. Integration of the chromosome by a specialized transducing phage involves a recombination event between the circular bacterial chromosome and the phage DNA. This recombination is site-specific and occurs at a specific attachment site which is called as core sequence. This core sequence is 15 bp in length and is identical in sequence in both *E. coli* chromosome and phage DNA. This site-specific recombination event results in the covalent linear insertion of the phage chromosome into the chromosome of the bacterium. The recombination occurs at specific attachment sites present in both phage and host chromosomes. The attachment sites of the coliphage and *E. coli* contain the following identical 15 bp sequence called the core sequence.

The integrated phage chromosome is called a prophage. The lytic genes of the virus replication are inhibited or repressed. A bacterium harbouring a prophage is called the lysogenic bacteria and the relationship is called as lysogeny. As lytic genes of prophage are repressed, lysogenic bacteria cannot be infected by secondary phage, especially by same virus (Figure 12.5). Interestingly a lysogeny state can be reverted back to lytic cycle or lysis by exposing to UV light. The excision of prophage from the host chromosome is very precise and site-specific. Occasionally, however, the excision event occurs at a site other than the original attachment site. Thus a portion of the bacterial chromosome is excised with the phage DNA, leading to the formation of specialized transducing particles transmitted to only one. Only host genes located close to the site of prophage insertion can be excised with the phage DNA.

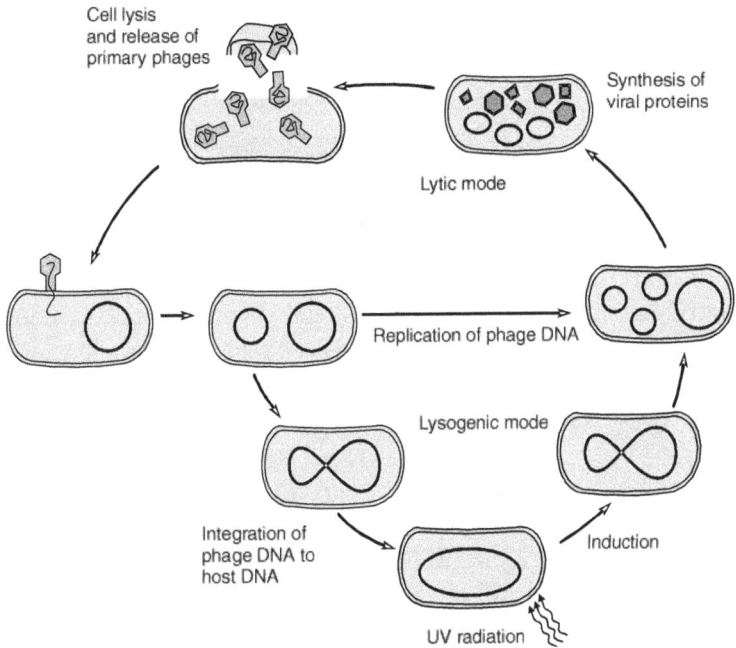

Figure 12.5 Lysogeny and lytic steps in bacteriophage infection

When a specialized transducing particle infects a host cell, the phage chromosome may integrate into the host chromosome through recombination at the attachment site. The host chromosome now contains the incomplete phage chromosome as well as the segment of host chromosome present in the transducing particle. Such a host cell is partially diploid, and is called as transductant or transduced cell. The phage DNA integrated into transductant may be excised with or without the host DNA. If it is excised without the host DNA, then the phage DNA is lost as it can reproduce by itself, whereas the integrated DNA, i.e., extra gene obtained from the other bacteria, will continue to exist in the new host.

Mapping of Chromosomes by Transduction

Transduction can be used to establish gene order and map distance. Gene order can be established by two-factor transduction. For example, if gene A is co-transduced with gene B and B with gene C but A is never co-transduced with C, we have established the order ABC. This would also apply to quantitative difference in co-transduction to detect cells that have been transduced for one, two or all the three loci. We need to eliminate the non-transduced cells. In other words after transduction, there will be A^-, B^-, C^- cells in which no transduction event has taken place. There will also be seven classes of bacteria that have been transduced for one, two or all three loci ($A^+B^+C^+$, $A^+B^+C^-$, $A^+B^-C^+$, $A^-B^+C^+$, $A^+B^-C^-$, $A^-B^+C^-$ and $A^-B^-C^+$). The simplest way to select for transduced bacteria is to select bacteria in which the wild type has replaced at least one of the loci. For example if after transduction, we grow the bacteria in minimal media the requirements of B^- and C^- added, all the bacteria that are A^+ will grow. Hence, although we lose the untransduced bacteria ($A^-B^-C^-$), replica-plating allows us to determine gene type at the B and C loci for the A^+ bacteria. In this example, colonies that are grown on complete medium without the requirement of the A mutant are replica-plated onto a complete medium without the requirement of the B mutant and then onto complete medium without the requirement of the C mutant. In this way, each transductant can be scored for the other two loci. Now let us take all these selected transductants in which the A^+ allele was incorporated. These can be of four categories $A^+B^+C^+$, $A^+B^+C^-$, $A^+B^-C^-$ and $A^+B^-C^-$. We can now compare the relative numbers of each of these four categories. The rarest category will be caused by the event that brings in the outer two markers, but not the centre one, because this event requires for crossovers. Thus, by looking at the number of transductants in the various categories, we can determine that the gene order is ABC.

In transformation, we measure the co-occurrence directly; therefore, the measure, co-transductance, is the inverse of map distance. In other words, the greater the co-transduction rate, the closer the two loci are; the more frequently two loci are transduced together, the closer they are and the higher the co-transduction value (Figure 12.6). From these sorts of transduction experiments,

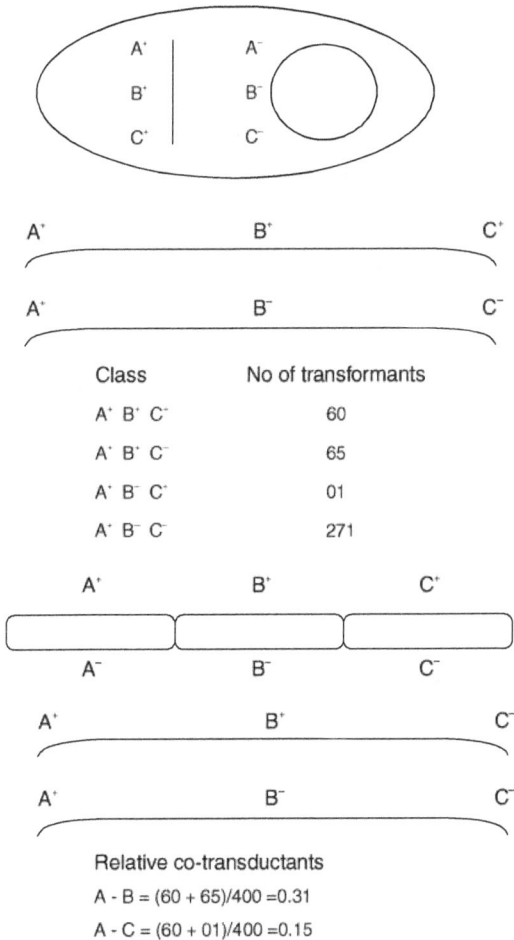

Class	No of transformants
A⁺ B⁺ C⁻	60
A⁺ B⁺ C⁻	65
A⁺ B⁻ C⁺	01
A⁺ B⁻ C⁻	271

Relative co-transductants

A - B = (60 + 65)/400 =0.31

A - C = (60 + 01)/400 =0.15

Figure 12.6 Diagram depicting the sequence of events that occur during co-transduction

it is possible to round out the details of map relations of *E. coli* after obtaining the overall picture by interrupted mating. Unlike the measurements in eukaryotic mapping, prokaryotic map distances are not generally thought of in map units. Rather, the general distance between locus is determined in minutes with co-transduction values used for loci that are very close to each other.

TRANSFORMATION

Transformation was first discovered by Griffith in 1928 in *Diplococcus pneumoniae*. But Avery, Macleod and McCarthy explored the exact phenomenon and principle in 1944. In transformation, naked DNA molecules are taken in by bacterial cells. These DNA molecules replace homologous segments of the chromosomes of recipient cells. The uptake of DNA molecule by recipient bacteria is an active, energy–requiring process. It does not involve passive entry of DNA molecules through permeable cell walls and membranes. Thus, transformation does not occur naturally in all species of bacteria. Only those species possessing the enzymatic machinery involved in the active uptake and recombination show the phenomenon of transformation. Most of the studies on transformation are carried out in *D. pneumoniae*, *B. subtilis* and *H. influenzae*. Even in these species, all cells in a given population are not capable of active uptake of DNA. Only competent cells, which possess the so-called "Competence factor" are capable of serving as recipients in transformation.

Transformation consists of five stages.

1.   Binding of dsDNA to the receptor sites on the surface of the cell. This binding is a reversible binding.

2.   Uptake of dsDNA, bound to the surface, into the cell.

3.   The dsDNA is converted into single-stranded DNA by nucleolytic degradation of one strand. Out of the two

strands, any strand can be degraded, it is only chance which decides which one has to be degraded.

4. The single strand of linear DNA is integrated into the chromosome of the recipient. The integration is facilitated by homologous recombination. However, this step is very specific for homologous DNA. It does not mean that it will not occur with heterologous DNA. It can occur with heterologous DNA, but at a lower frequency than that of homologous DNA.

5. The last step is segregation and phenotypic expression of the integrated donor gene.

The first three steps in transformation, binding, uptake and degradation of one strand of dsDNA, are not specific. This event can be carried out when a human DNA is used whereas the degradation of DNA is carried out by using a specific exonuclease, DNA translocase. DNA translocase degrades one strand of the DNA while it transports the other strand near to nucleus. This process occurs in *D. pneumoniae,* but does not occur in other bacteria, when the translocase enzyme is expressed. Moreover, considerable evidence suggests that these processes may vary in different species.

Astonishingly, even after 80 years of discovery of transfer by Griffith, the exact mechanism by which transformation occurs is still unknown. We just have a picture of the overall process of transformation.

Mapping by Transformation

The general idea of transformation mapping is to add DNA from a bacterial strain with known genotype to another strain also with known genotype, but with different alleles at two or more loci. We then look for incorporation of the donor alleles into the recipient strain of bacteria. The more often alleles from two loci are incorporated together into the host the closer these loci must be to each other. Thus, we can use an index of co-occurrence that is in inverse relationship to map distance. The larger the co-occurrence of alleles of two loci the closer the loci must be.

For example, a recipient strain of *B. subtilis* is auxotrophic for amino acids tyrosine (tyr A⁻) and cysteine (cys⁻). We are interested in how close these loci are on the bacterial chromosome. We thus isolate DNA from a prototrophic strain of bacteria (tyrA⁺ cys⁺). We add this donor DNA to the auxotrophic strain and allow time for transformation to take place. If the experiment is successful, and the loci are close enough together, then some of the recipient bacteria may incorporate donor DNA that has both donor alleles or one or the other donor allele. Thus some of the recipient cells will now have the tyr A⁺ and cys⁺ alleles, some will have just the donor tyr A⁺ allele, some will have just the cys⁺ allele, and the overwhelming majority will be of the untransformed auxotrophic genotype, tyr A⁻ cys. We thus need to count the transformed cells (Figure 12.7).

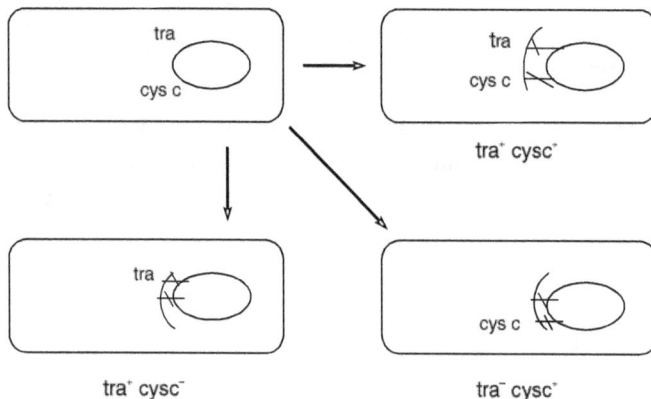

Figure 12.7   An example of mapping by transformation

After transformation, the cells are grown on a complete medium overnight. Then cells are replica-plated onto three plates—a minimal medium plate, a minimal medium plus tyrosine plate and a minimal medium plus cysteine plate and allowed to grow overnight in an incubator at 37°C. We then count the colonies. Those growing on minimal medium are of genotype tyr A$^+$ cys$^+$; those growing on minimal medium with tyrosine but not growing on minimal medium are tyr A$^-$ cys$^+$; and those growing on minimal medium with cysteine but not growing on minimal medium are tyrA$^+$cys$^-$. The overwhelming majority will grow on complete medium, but not on minimal medium or minimal media with just tyrosine or cysteine added. This majority is made up of non-transformants that is auxotrophs that were not involved in a transformation event. They took up no foreign DNA.

As a control against reversion, the normal mutation of tyr A$^-$ to tyr A$^+$ or cys$^-$ to cys$^+$, we grow several plates of auxotrophs in minimal medium and minimal medium with tyrosine or cysteine added. These are auxotrophs that were not exposed to prototrophic donor DNA. We then count the number of natural revertants and correct our experimental numbers by the natural reversion rate. Thus we are sure that what we measure is the actual transformation rate rather than just a mutation rate that we mistake for transformation. This control should always be carried out.

Assuming that in this experiment, we get 15 double transformants (tyrA$^+$, cys$^+$), 21 tyr A$^+$ cys$^+$ and 30 tyrA$^-$ cys$^+$ from these data. We calculate the co-occurrence or co-transfer index.

$$R = \frac{\text{Number of double transformants}}{\text{Number of double transformants} + \text{Number of single transformants}}$$

$$r = \frac{15}{(15+12+30)}$$

$$= \frac{15}{66} = 0.22$$

This is a relative number which indicates the occurrence of the two loci and thus their relative distance apart on the bacterial chromosome. By systematically examining many loci, we can establish their relative order (e.g. ABC) but we cannot determine exact order for very closely linked genes. Bacteria and viruses are ideal subjects for biochemical analysis and genetic analysis.

RELATIVE MERITS OF THE THREE METHODS

The existence of three different mapping techniques has brought about great advances in understanding gene organization in bacteria. To map a newly discovered gene in *E. coli* a good approach would be to first perform an interrupted mating experiment, which positions the gene with an accuracy of about 2 minutes, equivalent to about 92 kb of DNA. The exact position of the gene can then be established with greater accuracy by testing for co–transformation or co-transduction with genes known to lie in the same region of the genome, finally, the gene order can be worked out by co-transduction and co-transformation. This situation does not hold good for all bacteria. Conjugation mapping is clearly limited to those species that possess F-plasmid or equivalents. This means that it is not generally applicable to the gram-positive group of bacteria, which includes several important genera. Transduction mapping is also limited in scope because transducing phages are known for only a few species. For many bacteria, transformation mapping is the only technique that can be used satisfactorily.

These limitations are not so important today because recombination techniques have provided new gene mapping procedures that do not depend on gene transfer in the conventional sense and so are applicable to virtually all species including higher organisms. Recombinant DNA technology also enables the researcher to go several stages beyond gene mapping and obtain detailed information on gene structure and expression.

TETRAD ANALYSIS

For Drosophila and other diploid eukaryotes, genetic analysis is referred to as random strand analysis as zygotes are the result of the random uniting of chromatids. Generally fungi are haploid, the zygotes are products of fertilization of haploid cells, which undergo meiosis to produce haploid fungi. Hence there is no need for the test crosses so common in the case of diploid organisms.

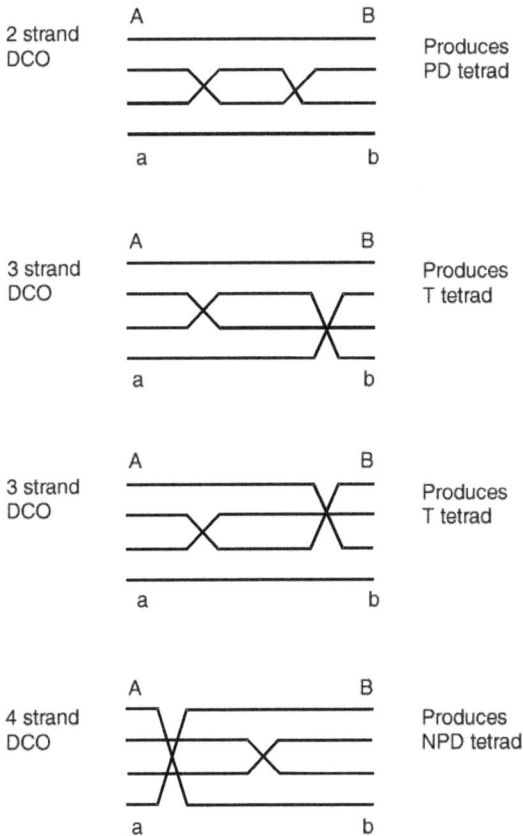

Figure 12.8 Recombination in chromosome

In many fungi, the spores produced due to different meiotic events are enclosed in separate membranes, e.g. asci of *Neurospora*. These structures are called "tetrads". Since each of them has the four haploid cells called tetrad, the result is from one complete meiotic division. In some fungi especially *Neurospora*, the products of different divisions during spore formation are arranged in the precise sequence in which they are produced. Such tetrads are called as ordered tetrads. Ordered tetrads help us in estimation of recombination frequency between a gene and its centromere.

In fungi, the detection of linkage and the estimation of recombination frequencies between genes is based on the classification of a large number of tetrads produced by the appropriate heterozygote into different classes, this is known as tetrad analysis. When an ab spore fuses with an a⁺b⁺ spore and the diploid then undergoes meiosis, the spores can be isolated and grown as haploid colonies, which are then observed for phenotypes that the two loci control. Only three patterns can occur. The first class has two types of spores which are identical to the parental haploid spores. This ascus type is referred to as a parental ditype (PD). The second class also has only two spore types, but they are recombinants. This ascus type is referred to as a non-parental ditype (NPD) (Figure 12.8). The third class has all four possible spore types and is referred to as a tetratype (TT). Of the ascomycetes, the bread mould *Neurospora crassa* has been of particular importance in genetic studies. Five features of *N. crassa* make it highly suited for certain types of genetic analysis.

1. After two haploid nuclei from cells of two different mating types fuse to form a diploid fusion nucleus, meiosis occurs, just as it does in higher plants.

2. The haploid products of meiosis, called ascospores, are maintained in a linear order within an elongate, tube like structure called ascus. Thus each ascus contains all four

products of single meiotic event. Moreover, all of the ascospores in each ascus (i.e., spore) can usually be recovered and analysed genetically.

3. These haploid ascospores germinate and grow to produce multicellular mycelia and all of these cells are also haploid. The genotype of each product of meiosis can thus be determined without carrying out test crosses or other genetic manipulations. Because of the haploid state of the mycelium, the presence of recessive markers is never masked by dominant alleles.

4. *Neurospora crassa* can grow on simple medium.

5. *Neurospora crassa* reproduces asexually as well as sexually, and facilitates the maintenance of strains of particular genotypes.

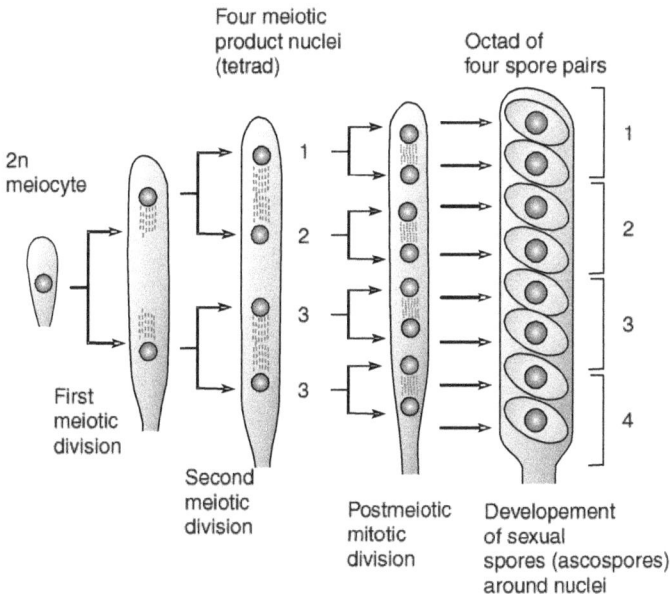

Figure 12.9 Tetrad formation in *Neurospora*

Since the *Neurospora* ascus is narrow, the meiotic spindle is forced to lie along the cell's long axis (Figure 12.9). The two nuclei then undergo the second meiotic division which is also oriented along the long axis of the ascus. The result is that the spores are ordered according to their centromeres. That if if we label one centromere "A" and other "a" for the two mating types, a tetrad at meiosis will consist of one "A" and one "a" centromere. At the end of meiosis in *Neurospora*, the four ascospores are in the order AAaa or aaAA with regard to centromere.

If there was no crossover between the locus and its centromere the allelic pattern is the same as the centromeric pattern, which is referred to as first division segregation (FDS), e.g. (AAAAaaaa). If however, a crossover has occurred between the locus and its centromere, patterns of a different type emerge, which is referred to as second division segregation (SDS), e.g. (AAaaAAaa). Because the spores are ordered, the centromeres always follow a first division segregation pattern. Hence, we should be able to map the distance of a locus to its centromeres, under the simplest circumstances, every second division segregation configuration has four recombinant and four non-recombinant chromatids. Thus, half of the (chr) spores in a second division segregation ascus are recombinant.

$$\text{Map distance} = \frac{1/2 \text{ of SDS asci}}{\text{Total number of asci}} \times 100$$

REVIEW QUESTIONS

1. How do you distinguish between conjugation, transformation and transduction?

2. Describe the structural difference between F⁺ and F⁻ strains or donor and recipient cells of bacteria.

3. Describe the molecular mechanism of conjugation.

4. Describe how the mechanism of conjugation can be used for mapping a chromosome.

5. Explain the phenomenon of sexduction.

6. Describe transduction in detail.

7. Explain the fundamental difference between generalized and specialized transduction.

8. Describe the relative merits of conjugation, transformation and transduction.

9. Explain tetrad analysis.

13

TRANSPOSONS

Genes evolve both by acquiring new sequences by mutations or by rearranging existing sequences. Another major cause of the variations is provided by "transposable elements" or "transposons".

Transposons are the segments of DNA able to move from one site to another in the genome or between genomes in the same cell. Transposons can move without the help of any independent factor like plasmids and they cannot transfer from one cell to the other like transformation or conjugation. They are independent and can move from one place to the other independent of sequence homology or similarity. Studies with diverse organisms including bacteria, fungi, insects, etc. suggested that transposable elements are widespread with considerable variations in structure and function at the molecular level.

BACTERIAL TRANSPOSONS

Bacterial transposons were the first to be studied at the molecular level and provided maximum information about the transposons architecture and function. Typically, they are quite small ranging from 500 to 10,000 bp. Bacteria contain transposons which belong to two main classes   1) IS elements and   2) Composite elements or in family transposons.

IS Elements

They are simple and were first discovered in bacteria. They are typically less than 1500 bp in length. There is a single coding sequence with short, identical or nearly identical sequences at both ends (Figure 13.1). These terminal sequences are always in inverted orientation with respect to each other and so are called inverted terminal repeats, e.g. if one end of an insertion sequence is 5′–AGTCCG 3′; the other end of that strand will be the reverse complement 5′ GCCTGA – 3′. The main body of an insertion sequence codes for at least two proteins that catalyse transposition. These proteins are called as transposases. When IS elements insert into chromosome, they create a duplication of the DNA sequence at the site of the element. These short, directly repeated sequences are therefore called target site duplications. These repeats did not exist before the transposons were inserted; they result from the insertion process itself and tell us that the transposase cuts the target DNA in a staggered fashion rather than with two cuts right across each other. The length of these direct repeats depends on the distance between the two cuts in the target DNA strands. This distance depends in turn on the nature of the insertion sequence, e.g. IS1 makes cuts nine base pairs apart and therefore generates direct repeats that are nine base pairs long. These target site duplication sequences act as footprints of transposon insertion and removal, as these sites do not form the integral part of transposon. Hence they are not transferred along with the transposon.

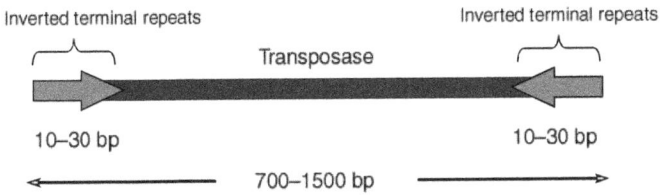

Figure 13.1   IS type transposon structure

Composite Transposons

Composite transposons are created when two IS elements get inserted near each other. A composite transposon consists of a central region surrounded by two IS elements (Figure 13.2). The central region usually contains transposase gene and bacterial gene, frequently the antibiotic resistance gene. The terminal repeats can be similar or different. Based upon the sequence and orientation of terminal repeats and type of antibiotic resistance, composite transposons are further divided into families called as T_n series.

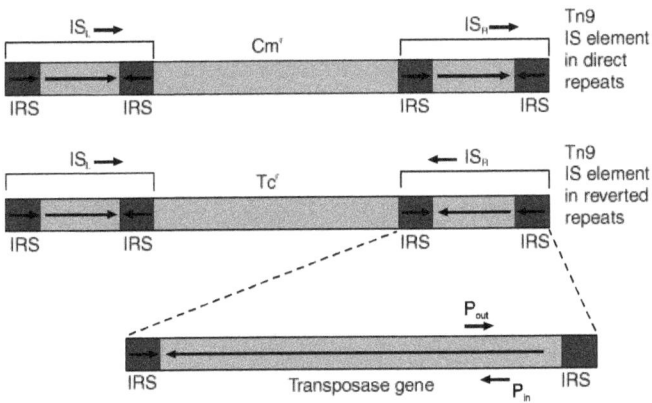

Figure 13.2   Molecular architecture of composite transposons

Tn9   Tn9 transposons carry chloramphenicol resistance gene and have inverted repeats which are 23 bp in length and are in direct orientation. The entire length of transposon is 2500 bp.

Tn5   It contains kanamycin resistance gene and has an inverted repeat sequence in inverted orientation and which is 9 bp in length. The total length of this transposon is 5700 bp in length. The inverted repeat at the left side is called as IS50L and right as IS50R, due to sequence difference (Figure 13.3).

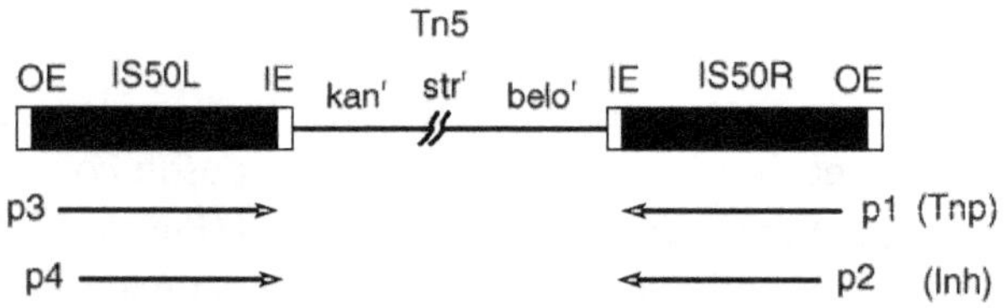

Tn5

| OE | IS50L | IE | kan' | str' | belo' | IE | IS50R | OE |

p3 ──────→   ←────── p1 (Tnp)
p4 ──────→   ←────── p2 (Inh)

Figure 13.3   Structure of Tn5 transposon

Tn10   It contains tetracycline resistance gene and has an inverted repeat sequence in inverted orientation and of length 22 bp. The total length of this transposon is 9300 bp.

Tn3   It is the widely studied transposon of composite family. The elements in this group of transposons have inverted repeats that are 38–40 bp long and produce a target duplication of 5 bp. There are three genes *tnpA*, *tmpA*, and *bla*. *bla* codes for β-lactamase, thus conferring resistance to the antibiotic ampicillin. *tnpA* codes for transposase whereas *tmpA* codes for resolvase, both of which play a role in transposition.

TRANSPOSABLE ELEMENTS IN EUKARYOTES

A detailed study regarding transposable elements has been done in bacteria, but a considerable amount of work is done and more yet to be done in eukaryotes, especially maize and yeast.

Yeast Ty Elements

Transposons of yeast are called as Ty elements and about 35 copies of transposable or Ty elements are present in a haploid genome. The transposons are about 5900 bp long and are bounded at each end by approximately 340 bp long DNA segments called "d" sequence or long terminal repeats or LTRs. Ty element is flanked by five base pairs which are direct repeats created by the duplication of DNA at the site of the Ty insertion. These target site duplications do not have a standard sequence, but they tend to contain A–T base pairs. This indicates a possibility that Ty elements preferentially insert into A–T-rich regions of the genome. Ty elements have only two genes, A and B, which are analogous to the *gag* and *pol* genes of the retroviruses. Biochemical studies have shown that the products of these two genes can form virus particles inside the yeast cells. However it is not known whether these particles are genuinely infectious. One hypothesis is that yeast Ty elements are primitive retroviruses, capable of moving between cells. In this regard, it has been shown that the transposition of Ty elements involves RNA intermediate. After the RNA is synthesized from Ty DNA, a product of the TyB gene uses the RNA to make double-stranded DNA. This process is called reverse transcription. Then the newly synthesized DNA is inserted somewhere in the genome, creating a new Ty element. Because of their overall similarity to the retroviruses, Ty elements are sometimes called retrotransposons.

Maize Transposons

The first transposon that was discovered was maize transposon elements by Barbara McClintock. The Ac and Ds elements are prominently found transposons in maize.

Ac element   Ac element is 4563 base pairs long and bounded by 8 base pair terminal repeats. This direct repeat is created at the time that the elements insert into a site on a chromosome. Other repeat sequences are found within the element itself. The most conspicuous of these are at the ends, where an 11-nucleotide pair sequence at one end is repeated in the opposite orientation at the other end. All the Ac elements in the maize genome appear to be structurally similar.

The activating function of Ac element is associated with a protein that it synthesizes. Since this protein is involved in transposition, it is sometimes called as transposase of the Ac family. Deletions or mutations in the gene that encodes this protein abolish the activating signal, thus they cannot activate themselves. However, a single Ac element can provide it to all the Ac and Ds elements in the genome. Recently, biochemical analysis has indicated that the activity of Ac element is controlled by the methylation of the selected nucleotides in the DNA sequence.

Ds elements   Ds elements are formed from the Ac elements by deletions, additions or rearrangements. Ds elements show a considerable structural heterogeneity. One type or class of Ds elements is derived from Ac elements by deletions of internal sequences. In another type, one Ds element is inserted into another, but in an inverted orientation. It has been shown that these double Ds elements are responsible for chromosome breakage.

Retro Elements

These are generally transposable elements which mobilize by reverse transcribing an RNA intermediate and then integrating cDNA into the genome. Retro elements are absent in prokaryotes, but widely found in yeast and vertebrates. There are two types of retro elements.

These elements have structural features similar to retroviruses, long terminal repeats each flanked by short inverted repeats and a central region containing several open reading frames (Figure 13.4). The central region of retrotransposons contains open reading frames homologous to *gag* and *pol*. The latter exceeding reverse transcriptase and integrase, *env* gene is either disrupted or deleted. The mechanism of retrotransposon activity is otherwise similar to that of retroviruses. Some argue that retro elements are inactivated retroviruses. These are also called as class I.1 transposable elements.

a. Viral retrotransposons

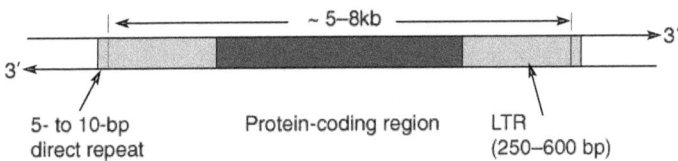

5- to 10-bp direct repeat

Protein-coding region

LTR (250–600 bp)

b. Genes in retrorival DNA and viral retrotransposons

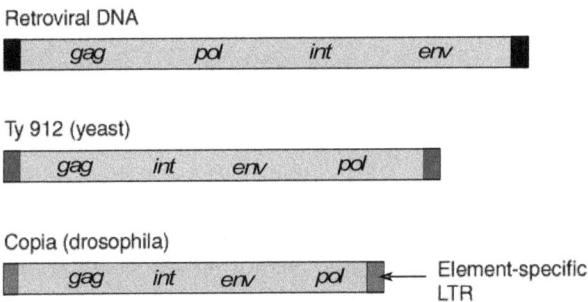

Retroviral DNA

gag   pol   int   env

Ty 912 (yeast)

gag   int   env   pol

Copia (drosophila)

gag   int   env   pol   Element-specific LTR

Figure 13.4   Retrotransposon molecular structure

Retroposons

These elements represent all the retro elements that do not resemble retroviruses. These elements do not have long terminal repeats, but contain the central region homologous to *gag* and *pol* genes. These elements generate target site duplications varying in size. These retro elements are also called as class I.2 transposable elements.

MECHANISM OF TRANSPOSITION

The process of transfer of a transposon from one place to another is called as transposition. McClintock coined the term transposition. There are three steps in transposition. First the transposable element is separated from the host DNA by an endonucleolytic cleavage reaction with the help of transposase enzyme. This step is called as donor cut stage. Secondly, at the target site, the transposase enzyme inserts the transposable element by transesterification reaction. This step is called as strand exchange stage. In the last step, which is called as repair stage, DNA polymerase will fill any gap that is generated by transposon insertion. Transposable elements vary in their preference for target sites, some integrating at specific sequences, while others integrate at random. Transposons, which show site preferences must encode resolvase, a site-specific recombination, which acts on a short sequence within the transposon, the internal resolution site (res). Resolvase exhibits strong directionality in its activity.

Transposition mechanism can be either conservative or replicative. In conservative transposition, transposon is excised from one site and is integrated at another site. It does not increase the copy number of the element. It is also called as non-replicative or simple transposition or cut-and-paste mechanism (Figure 13.5).

In replicative transposition, the donor transposable element undergoes replication (not whole DNA replication) and the copy is integrated at the target site. In replicative transposition only two steps (strand exchange and repair stages) are present.

If one transposon undergoes mutation and loses its ability to produce transposase enzyme, still it can move from one place to the other by using enzymes coded by non-mutated transposable elements. Such transposons are called as non-autonomous elements. The process where the non-autonomous elements move with the help of enzymes obtained in trans is called as passive transposition.

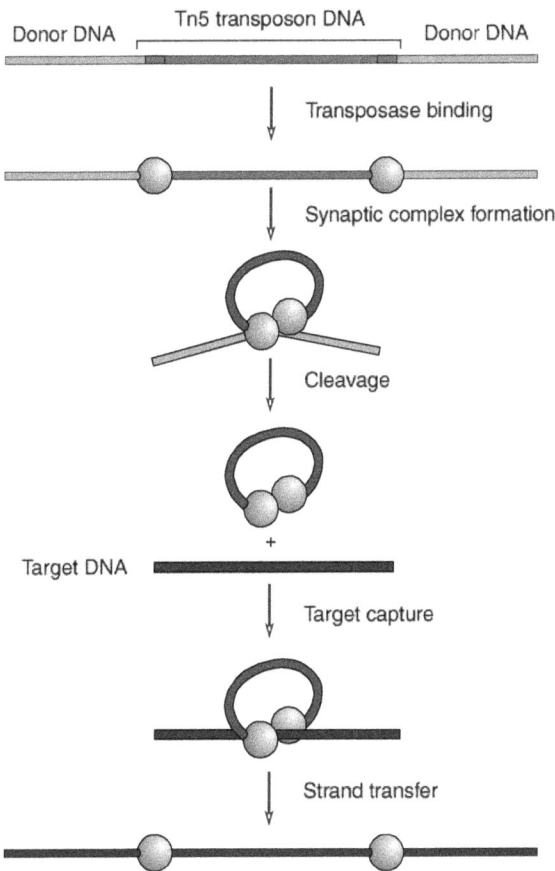

Figure 13.5 Mechanism of transposition by a non-replicative type of transposon

SIGNIFICANCE OF TRANSPOSABLE ELEMENTS

1. Transposable elements are responsible for mutations in a wide variety of organisms. Transposable elements also produce chromosome breakage that leads to the loss or rearrangement of chromosomal material.

2. Bacterial transposons are clearly responsible for the transposition of genes controlling resistance to antibiotics from one molecule to another.

3. In several organisms, it is now feasible to stimulate the transposition of a particular family of elements, thereby increasing the natural mutation rate. This procedure has an advantage over traditional methods of inducing mutations because a transposable element that has caused a mutation by inserting into a gene can serve as a landmark for more detailed studies.

4. The widespread distribution of transposable elements suggests that they have played a role in evolution. These elements are natural tools for genetic engineering. Their ability to copy, transport and rearrange other DNA sequences is beneficial to the organisms that carry them.

5. Their ability to jump between genomes has enabled them to be used as gene delivery vectors and their tendency to disrupt genes has allowed them to be exploited as mutagens. They have been used to introduce genome rearrangements, as mobile reporter systems and as tags to clone nearby genes.

REVIEW QUESTIONS

1. Explain in detail the structure of bacterial transposon.

2. Describe the mechanism of transposition with an example.

3. Write in detail about eukaryotic transposons.

4. Describe the significance of transposable elements.

5. Write short notes on the following.

   i. Tn elements

   ii. Composite transposons

   iii. Ds elements

   iv. Transposons in maize

   v. IS elements

   vi. Retro elements

   vii. Yeast Ty elements

   viii. Replicative transposons

   ix. Non-replicative transposons

14

CHLOROPLAST AND MITOCHONDRIAL GENOME ORGANIZATION

Till now, we have dealt with genes present in the nuclei of eukaryotes. Certainly, nuclear DNA is the most important and very nearly the universal genetic material. But there are evidences for the presence of genes outside the nucleus.

Till now, we have discovered that both the chloroplast and mitochondria have DNA of their own. These DNA inherit independently of nuclear genes. In effect, the organelle genome comprises a length of DNA that has been localized in a defined part of the cell and is subject to its own form of expression and regulation. An organelle genome can code for some or all of the RNAs, but codes for only some proteins are needed to perpetuate the organelle. The other proteins are coded in the nucleus, expressed via the cytoplasmic protein synthetic apparatus. Genes not residing within the nucleus are generally described as extranuclear genes.

MITOCHONDRIAL DNA ORGANIZATION

The mitochondrion is an organelle in a eukaryotic cell in which the electron transport chain takes place. The actual number of mitochondria per cell can be determined by electron microscopy. The most interesting aspect of the mitochondrion is that it has its own DNA. Mitochondria provide higher animals and plants with life-sustaining cellular energy through the oxidative processes of the citric acid and fatty acid cycles.

Animal mitochondrial DNA is extremely compact, with very few non-coding regions and no introns. Each strand of duplex is transcribed into a single RNA product that is then cut into smaller pieces primarily by freeing the twenty-two transfer RNAs interspersed throughout the genome. Also formed are a 16S and a 12S ribosomal RNA. Although proteins and small molecules such as ATP and tRNAs can move in and out of the mitochondrion, large RNAs cannot.

Most interestingly, all mtDNAs exhibit the same basic organization of genetic information. Each contains 2 rRNA genes, 22 tRNA genes, and 13 putative protein structural genes. Five genes encode known proteins but the products and functions of the other putative genes have not yet been identified. The entire mammalian mitochondrial genome is transcribed as one unit from a single promoter site and the giant primary transcript is then cleaved endonucleolytically to produce the individual tRNA, rRNA and mRNA molecule. Thus, the entire mtDNA is, in effect, equivalent to one operon in bacteria.

The human mitochondrion has 13 genes. But some proteins are transported into the mitochondrion as they are synthesized in the cytoplasm under the control of nuclear genes. Proteins targeting the entry into the mitochondrion have a signal peptide of 85 amino acids in length. These transit peptides are usually cleaved off the precursor polypeptides during their transport across the mitochondrial membrane.

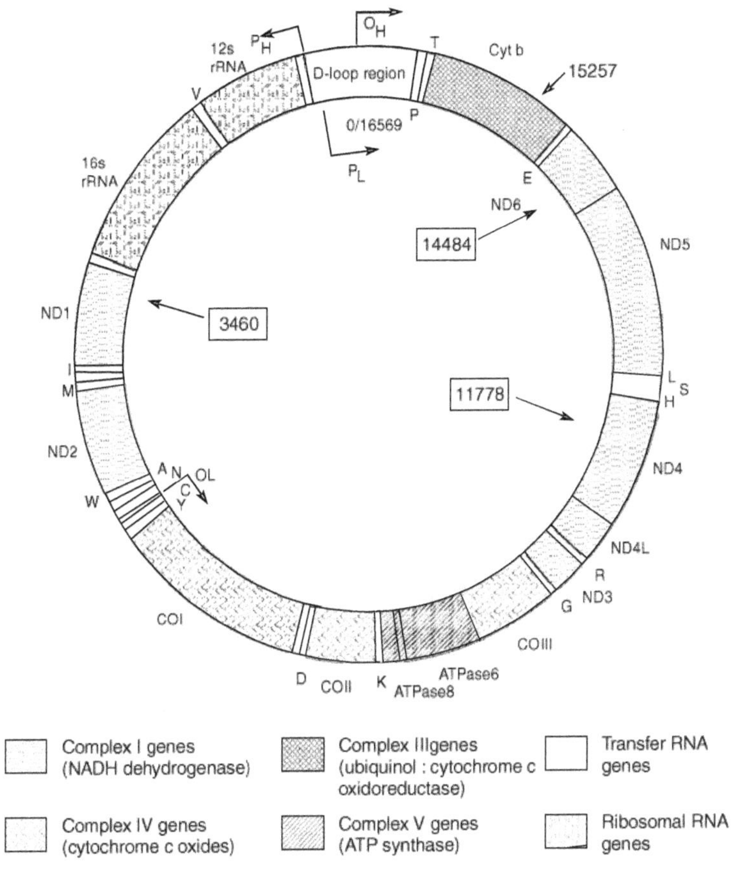

Figure 14.1 Mitochondrial genome organization

In mammalian mitochondrial genes, there are no introns. Some genes actually overlap and almost every single base pair can be assigned to a gene, with the exception of the D-Loop, a region concerned with the initiation of DNA replication (Figure 14.1).

In yeast cells, 10–20 percent of the cellular DNA is localized in a single mitochondrion. Yeast mitochondria have introns and are not economical like that of mammalian mitochondria. Some suggest that the genes with introns have been originated in the nucleus and are later captured or integrated by the mitochondria. Even the genome size is five times bigger than mammalian mtDNA.

CHLOROPLAST GENOME ORGANIZATION

Chloroplast is the chlorophyll-containing organelle that carries out photosynthesis and starch grain formation in plants. Like mitochondria, chloroplasts contain DNA and ribosomes, both with prokaryotic affinities. The DNA of chloroplast (cpDNA) is a circle that ranges in size from 120–190 kb (Figure 14.2). The chloroplast DNA, like mitochondrial DNA, controls the production of tRNAs, ribosomal RNAs and some of the proteins found within the organelle.

In more than 25 chloroplast DNAs from different plant, species that have been sequenced, there seems to be about 87–183 genes in the chloroplast genome. The chloroplast genome codes for all the rRNA and tRNA species needed for protein synthesis. The ribosome includes two small rRNAs in addition to the major species. The tRNA set may include all of the necessary genes. The organelle genes are transcribed and translated by the apparatus of the organelle.

About half of the chloroplast genes codes for proteins involved in protein synthesis. The endosymbiotic origin of the chloroplast is emphasized by the relationship between these genes and their counterparts in bacteria. The organization of the rRNA genes in particular is closely related to that of a cyanobacterium, which pins down more precisely the last common ancestor between chloroplast and bacteria.

Introns in chloroplast fall into two general classes. Those in tRNA genes are usually located in the anticodon loops, like the introns found in yeast nuclear tRNA genes. Those in protein-coding genes resemble the introns of mitochondrial genes. This places the endosymbiotic event at a time in evolution before the separation of prokaryotes with uninterrupted genes.

The role of the chloroplast is to undertake photosynthesis. Many of the chloroplast genes code for proteins that are located or function in thylakoid membranes. But quite strangely, even some protein complex found on thylakoid membrane are coded by nuclear genes. Thus on the thylakoid membranes you find proteins coded both by chloroplast genome and nuclear genome. Other chloroplast complexes are coded entirely by one genome.

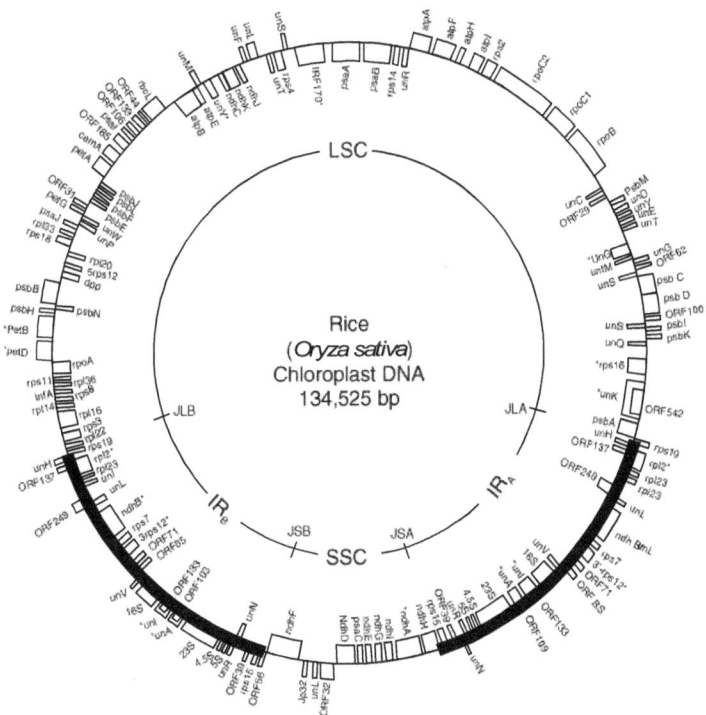

Figure 14.2   Chloroplast genome organization in rice

ORIGIN OF CYTOPLASMIC ORGANELLES

Extranuclear inheritance associated with cytoplasmic organelles meets the preceding requirements and deserves an explanation, although organelles represent only a very small proportion of genetic material, perhaps a few hundred genes, based on the amount of functional DNA that is involved. The fascinating possibility suggested by several earlier investigators and recently elaborated by Margolis is that mitochondria/chloroplasts were once free-living bacteria. Over long periods of time, they established a hereditary symbiosis with their eukaryotic host cells and ultimately evolved into organelles with animal and plant cells.

The first cells formed lacked mitochondria and chloroplast. Probably within a short time after the first eukaryotes appeared one of them engulfed a bacterium. However, instead of treating the bacterial cell as food and digesting it, the larger cell tolerated its guest. The two cells divided more or less together, producing many large progeny cells containing smaller guests. In time, the smaller cells began to specialize in energy production, but they lost many genes, so they could no longer live independent of their host. At the same time the host lost some of its genes so it could no longer produce energy efficiently; it became dependent on its guest for energy. Several lines of evidence point to this symbiotic association. Transcription in organelles is inhibited by the antibiotic rifampicin, which inhibits prokaryotic, but not eukaryotic RNA Pol. Even the subunits of RNA polymerase are similar to prokaryotes. Promoters for gene expression and protein synthesis in organelles also shows 100% similarity to that of prokaryotes. The size and sequence of rRNA genes and ribosomes are more closely related to prokaryotic genes rather than eukaryotic genes.

REVIEW QUESTIONS

1. Describe the similarities between the chloroplast, mitochondria and bacteria.

2. Explain the origin of cytoplasmic organelles with examples.

PART III

GENETIC ENGINEERING

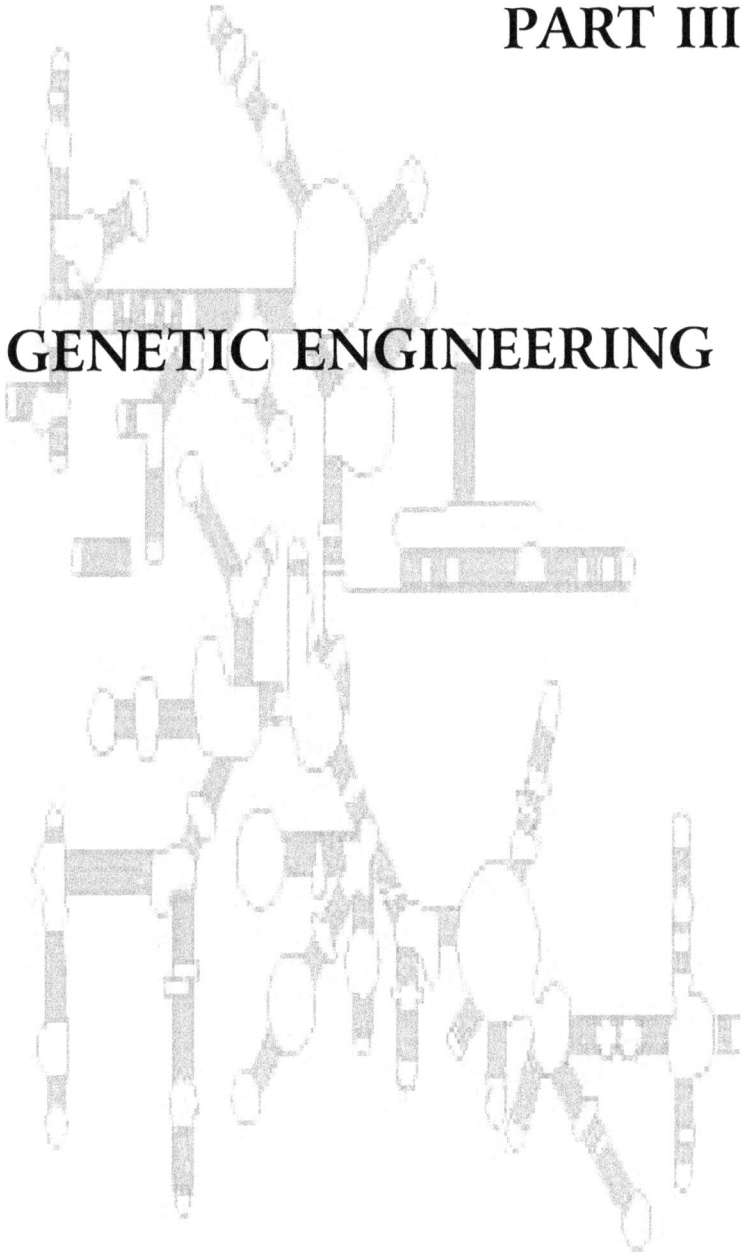

15

GENE CLONING

If a gene from humans is placed in bacteria, it does not produce any protein or multiply, as bacteria does not recognize the gene as a gene. For the recognition and multiplication of the gene it has to carry some identification sequences or replicons. Such replicons are known as vectors or cloning vehicles. A composite DNA molecule formed by joining a gene or insert with a cloning vehicle is called as recombinant DNA or chimeric DNA or chimeras. The process of creating a chimeric DNA is called as gene cloning, genetic engineering, gene manipulation or molecular cloning.

Every gene cloning experiment involves four basic steps (Figure 15.1).

1. The first step in gene cloning is isolation of DNA from the organism being studied. Sometimes the DNA is made by using mRNA as reference or by chemical synthesis (PCR). Whatever be the method, the first step is to obtain the DNA molecule of interest.

2. The next step is to construct the chimeric DNA or recombinant DNA molecules by joining the vector or cloning vehicle with insert. Depending upon the host cell

into which we transfer the insert, the vector or cloning vehicle varies. We use plasmids or phages for cloning in bacteria and SV40 virus for cloning in animal cells.

3. The next step is transfer of recombinant molecules into the host so that it multiplies in the host. When the host cell divides, copies of the recombinant DNA molecules are passed on to the progeny.

4. The last step is identification of the bacteria which carries a recombinant molecule, from an array of bacteria with no recombinant molecules or having only a cloning vector.

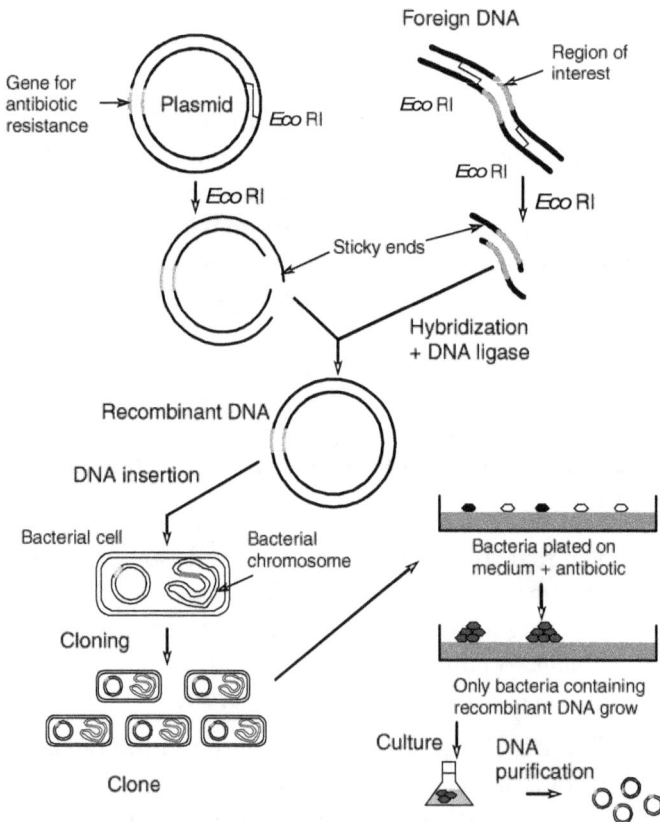

Figure 15.1   Steps involved in cloning of the gene into plasmid

After these initial steps, the protocols or procedures will diverge depending upon the goal of researchers. If they want to study the sequence of DNA, they proceed with DNA sequencing. If they want to make the protein of the gene they proceed for gene expression. Whatever be the ultimate goal, the abovesaid four steps remain constant and have to be carried out irrespective of whether the genetic engineer/researcher is fresh or experienced.

REVIEW QUESTION

1. Explain the basic steps in gene cloning.

16

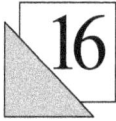

ENZYMES USED IN GENETIC ENGINEERING

Some technological developments in science enable us to gain immense knowledge and increased potential for innovation. Genetic engineering and biomedical research have experienced such a revolutionary change since the past 30 years with the development of gene manipulation. The ability to manipulate DNA *in vitro* (outside the cell) depends entirely on the availability of purified enzymes that can cleave, modify and join the DNA molecule in specific ways. At present, no purely chemical method can achieve the ability to manipulate the DNA *in vitro* in a predictable way. Only enzymes are able to carry out the function of manipulating the DNA. Each enzyme has a vital role to play in the process of genetic engineering.

NUCLEASES

Nucleases are a group of enzymes which cleave or cut the genetic material (DNA or RNA).

DNase and RNase

Nucleases are further classified into two types based upon the substrate on which they act. Nucleases which act on or cut the DNA are classified as DNases, whereas those which act on the RNA are called as RNases.

DNases are further classified into two types based upon the position where they act (Figure 16.1). DNases that act on the ends or terminal regions of DNA are called as exonucleases and those that act at a non-specific region in the centre of the DNA are called as endonucleases. Exonucleases require a DNA strand with at least two 5′ and 3′ ends. They cannot act on DNA which is circular. Endonucleases can act on circular DNA and do not require any free DNA ends (i.e., 5′ or 3′ end). Exonucleases release nucleotides (Nucleic acid + sugar + phosphate), whereas endonucleases release short segments of DNA.

An endonuclease

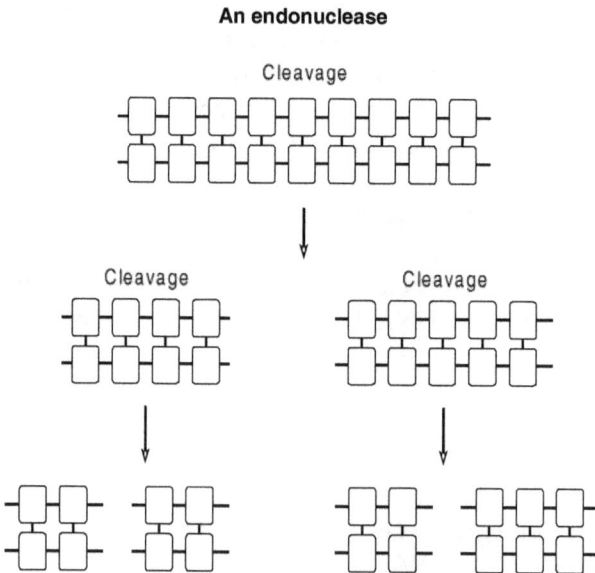

An exonuclease

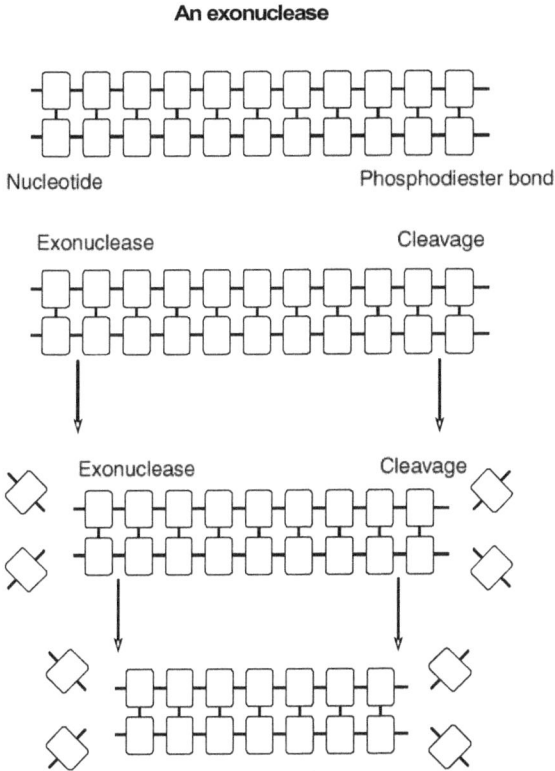

Figure 16.1 Diagrammatic representation of the difference between the site and products generated by an endo- and exonuclease

RESTRICTION ENZYMES

DNases which act on specific positions or sequences on the DNA are called as restriction endonucleases. The sequences which are recognized by the restriction endonucleases or restriction enzymes (RE) are called as recognition sequences or restriction sites. These sequences are palindromic sequences (i.e., if read from right to left, the sequence is same, e.g. MADAM). Different restriction enzymes present in different bacteria can recognize different or same restriction sites. But they will cut at two different points within the restriction site. Such restriction enzymes are called as *isoschizomers*. Interestingly no two restriction enzymes from a single bacterium will cut at the same restriction site.

Restriction enzymes are further classified into three types based upon the site of cleavage.

Type I These restriction enzymes recognize the recognition site, but cleave the DNA somewhere between 400 base pairs (bp) to 10,000 bp or 10 kbp right or left. The cleavage site is not specific. These enzymes are made up of three peptides with multiple functions. These enzymes require Mg^{++}, ATP and S-adenosyl methionine for cleavage or for enzymatic hydrolysis of DNA. These enzymes are studied for general interest rather than as useful tools for genetic engineering.

Type II Restriction enzymes of this type recognize the restriction site and cleave the DNA within the recognition site or sequence. These enzymes require Mg^{++} as cofactor for cleavage activity and can generate $5'-PO_4$ or $3'-OH$. Enzymes of this type are highly important because of their specificity. Type II restriction enzymes are further divided into two types based upon their mode of cutting.

Blunt-end cutters Type II restriction enzymes of this class cut the DNA strands at same points on both the strands of DNA within the recognition sequence (Figure 16.2). The DNA strands generated are completely base paired. Such fragments are called as blunt-ended or flush-ended fragments.

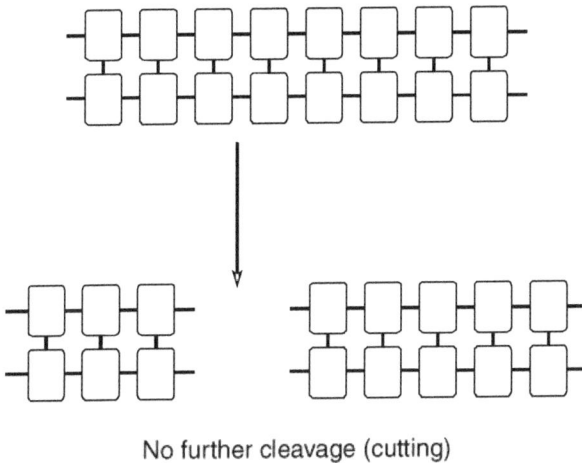

No further cleavage (cutting)

Figure 16.2 Blunt-end cutter

Cohesive-end cutter Type II restriction enzymes of this class cut the DNA stands at different points on both the strands of DNA within the recognition sequence. They generate a short single-stranded sequence at the end. This short single strand sequence is called as sticky or cohesive end. This cohesive end may contain $5'$-PO_4 or $3'$-OH, based upon the terminal molecule ($5'$-PO_4 or $3'$-OH). These enzymes are further classified as $5'$-end cutter (if $5'$-PO_4 is present) or $3'$-end cutter (if $3'$-OH is present).

Type III Restriction enzymes of this type recognize the recognition site, but cut the DNA 1 kbp away from the restriction site. These enzymes are made up of two peptides or subunits. These enzymes require ATP, Mg^{++} and S-adenosyl methionine for action.

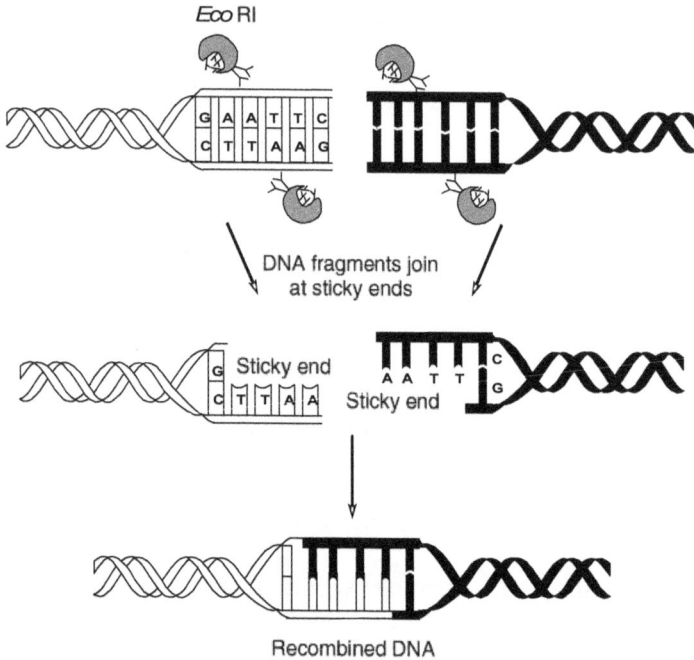

Figure 16.3   Generation of recombinant DNA

Nomenclature of Restriction Enzymes

As a large number of restriction enzymes have been discovered, a uniform nomenclature system is adopted to avoid confusion. This nomenclature was first proposed by Smith and Nattens in 1973.

1. The first letter of the restriction enzymes (RE) should be from the first letter of the species name of the organism

from which the enzyme is isolated. The letter should be written in capitals and italics, e.g. RE from *E. coli* will have *E* as starting letter.

2. The second and the third letters of RE should be from the first and the second letter of the genus name of the organism. The letter should be written in lower case and should be in italics, e.g. RE from *E. coli* will have *Eco* as starting words.

3. If the RE is isolated from the particular strain of an organism then that should be written as fourth letter. It should be in capitals and not in italics. For example, RE from *E. coli* R strain will be written as '*Eco* R'.

4. If the RE isolated is the first of its kind from that particular organism, then the number I should be given. If already two REs are isolated, then number III should be given for the new restriction enzymes. The number should be written in roman, e.g. the first *E. coli* RE should be written as *Eco* RI whereas the third restriction enzyme isolated from *E. coli* R strains should be written as '*Eco* RIII'.

Mode of Action of Restriction Enzymes

The restriction enzyme binds to the recognition site and checks for the methylation (presence of methyl group on the DNA at a specific nucleotide). If there is methylation in the recognition sequence, then, it just falls off the DNA and does not cut. If only one strand in the DNA molecule is methylated in the recognition sequence and the other strand is not methylated, then RE (only type I and type III) will methylate the other strand at the required position. The methyl group is taken by the RE from S-adenosyl methionine by using modification site present in the restriction enzymes. However, type II restriction enzymes take the help of another enzyme called methylase, and methylate the DNA. Then RE clears the DNA. If there is no methylation on both the strands of DNA, then RE cleaves the DNA. It is only by this methylation mechanism that, RE, although present in bacteria, does not cleave the bacterial DNA but cleaves the foreign DNA. But there are some restriction enzymes which function exactly in reverse mode. They cut the DNA if it is a methylate.

Star activity   Sometimes restriction enzymes recognize and cleave the DNA strand at the recognition site with asymmetrical palindromic sequence, for example *Bam* HI cuts at the sequence GATCC, but under extreme conditions such as low ionic strength it will cleave in any of the following sequence NGATCC, GPOATCC, GGNTCC. Such an activity of the RE is called star activity.

DNA LIGASE

Recombinant DNA experiments require the joining of two different DNA segments or fragments *in vitro*. The cohesive ends generated by some RE will anneal (join) themselves by forming hydrogen bonds. But the segments annealed thus are weak and do not withstand experimental conditions. To get a stable joining, the DNA should be joined by using an enzyme called ligase. DNA ligase joins the DNA molecule covalently by catalysing the formation of phosphodiester bonds between adjacent nucleotides. DNA ligase isolated from *E. coli* and T4 bacteriophage is widely used in molecular biology experiments. These ligases more or less catalyse the reaction in the same way, and differ only in requirements of cofactor. T4 ligase requires ATP as cofactor and *E. coli* ligase requires NADP as cofactor.

The cofactor is first split (ATP → AMP + 2Pi) and then AMP binds to the enzyme to form the enzyme–AMP complex. This complex then binds to the nick or break (with $5'$-PO_4 and

3′-OH) and makes a covalent bond in the phosphodiester chain (Figure 16.4). The ligase reaction is carried out at 4°C for better results.

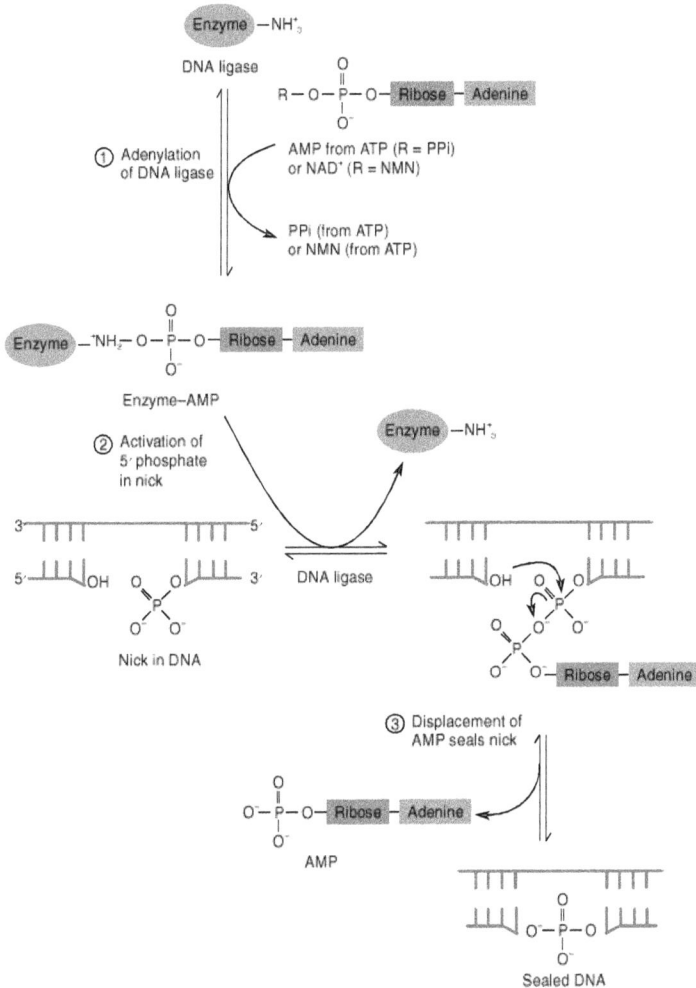

Figure 16.4 Chemical reactions that occur during ligase reaction

ALKALINE PHOSPHATASES

Phosphatases are a group of enzymes which remove a phosphate from a variety of substrates like DNA, RNA and proteins. Phosphatases which acts in basic buffers with pH 8 or 9 are called as alkaline phosphatases. Most commonly bacterial alkaline phosphatases (BAP), calf intestine alkaline phosphatases (CIAP) and shrimp alkaline phosphatases are used in molecular cloning experiments. The PO_4 from the substrate is removed by forming phosphorylated serine intermediate Figure 16.5. Alkaline phosphatase are metalloenzymes, and have Zn^{++} ions in them. BPA (bacterial alkaline phosphatase) is a dimer containing six Zn^{++} ions, two of which are essential for enzymatic activity. BPA is very stable and is not inactivated by heat and detergent. Calf intestine alkaline phosphatases (CIAP) is also a dimer. It requires Zn^{++} and Mg^{++} ions for action. CIAP is inactivated by heating at 70°C for twenty min. or in the presence of 10 mM EGTA.

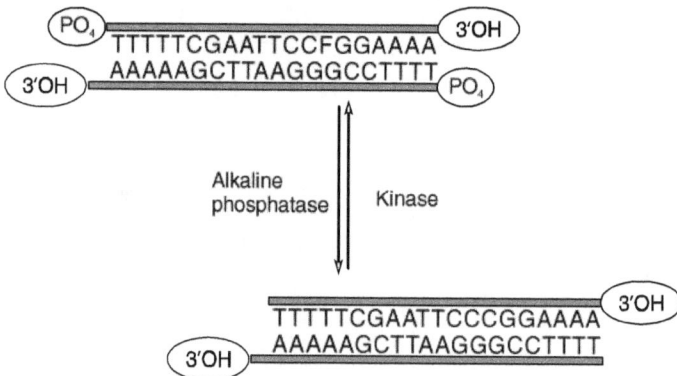

Figure 16.5   Function of kinase and alkaline phosphatases

Alkaline phosphatases are used to remove the PO_4 from the DNA or as reporter enzymes.

KINASE

Kinase is the group of enzymes, which adds a free pyrophosphate (PO_4) to a wide variety of substrates like proteins, DNA and RNA. It uses ATP as cofactor and adds a phosphate by breaking the ATP into ADP and pyrophosphate. It is widely used in molecular biology and genetic engineering to add radiolabelled phosphates.

KLENOW FRAGMENT

E. coli DNA polymerase I consists of a single polypeptide chain. *Pol I* carries out three enzymatic reactions that are performed by three distinct functional domains. Two fragments are obtained when DNA pol I is treated with trypsin/subtilisin in mild conditions. The larger fragment is called as klenow fragment. This fragment is 602 amino acids in length. The function of the klenow fragment is to add nucleotides to the 3´ end and 3´–5´ exonuclease activity.

Klenow fragment adds nucleotides by using complementary strand as reference. It cannot extend the DNA without the presence of the complementary strand. If any nucleotide is added by mistake and the base pair is wrong (i.e., if A is paired to G instead of T) then by using 3´–5´ exonuclease activity present in klenow fragment, this mispaired base pair is removed. In general the klenow fragment has 5´–3´ polymerase and exonuclease activity.

REVERSE TRANSCRIPTASE

This enzyme uses an RNA molecule as template and synthesizes a DNA strand complementary to the RNA molecule. These enzymes are used to synthesize the DNA from RNA. These enzymes are present in most of the RNA tumour viruses and retroviruses. Reverse transcriptase enzyme is also called as RNA-dependent DNA polymerase. Reverse transcriptase enzyme, after synthesizing the complementary strand at the 3′ end of the DNA strand, adds a small extra nucleotide stretch without complementary sequence. This short stretch is called as R-loop.

TERMINAL DEOXYNUCLEOTIDE TRANSFERASE

Terminal deoxynucleotide transferase is a polymerase which adds nucleotides at 3′-OH end (like klenow fragment) but does not require any complementary sequence and does not copy any DNA sequence (unlike klenow fragment). Terminal deoxynucleotide transferase (TDNT) adds nucleotide whatever comes into its active site and it does not show any preference for any nucleotide (Figure 16.6).

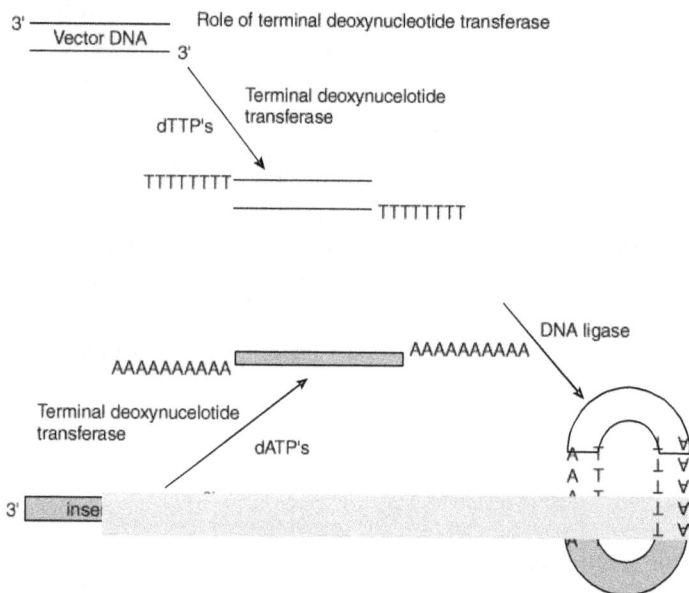

Figure 16.6　Terminal deoxynucleotide transferase action

RNase P

It specifically cleaves at the 5′ end of RNA. It is a complex
enzyme consisting of small protein (20 kilodaltons) and a

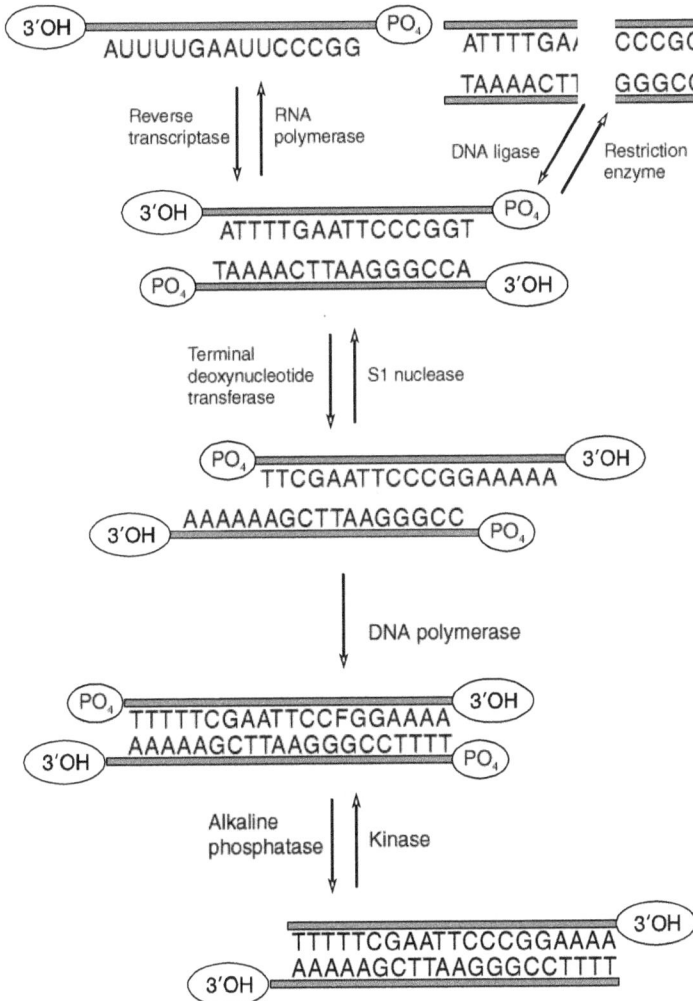

Figure 16.7 Block diagram showing functions carried out by
different enzymes

377-nucleotide RNA molecule. It has been observed that the RNA molecule possesses at least part of the enzymatic activity of the complex. Hence, it is an example of ribozyme.

The functions of the various enzymes used in genetic engineering are depicted in Figure 16.7.

REVIEW QUESTIONS

1. Write in detail about nuclease.

2. Explain the difference between exonuclease and endonuclease.

3. What is the difference between blunt-end and cohesive-end cutter?

4. Explain the mode of action of restriction enzyme and ligase used in genetic engineering.

5. Explain about different types of alkaline phosphatases used in genetic engineering.

6. Write the difference between alkaline phosphatases and kinase.

7. Write the difference between terminal deoxynucleotide transferase and klenow fragment.

8. Write short notes on the following.

    i.   Nomenclature of restriction enzymes

    ii.   Mode of action of restriction enzymes

    iii. Alkaline phosphatase

    iv. Kinase

    v.   Klenow fragment

17

BACTERIAL VECTORS

INTRODUCTION

Normally, when we mix bacteria or any cell with DNA or a gene of interest, the DNA or gene of interest is subjected to restriction enzyme or DNase digestion by the host because the host cell recognizes the DNA as foreign genetic material and assumes it as a virus or some infectious material. Hence for transferring a DNA or gene of interest into a bacteria or cell, we require a vehicle which carries our gene of interest and some identification tags which the bacteria recognizes as friend. Normally these identification tags are nothing but some specific DNA molecules which are unique for that bacteria or cell. These DNA molecules, which carry these identification tags are called as vectors in genetic engineering. There are many types of vectors, but the most common types are plasmids and cosmids.

PLASMIDS

Plasmids are autonomously replicating, extrachromosomal, covalently closed circular DNA molecules, which are present in prokaryotes. A plasmid can replicate when the chromosome is not in replication mode and when it is synthesizing proteins. As the chromosome and plasmids are not under the control of each other, plasmids are called as autonomously replicating, extrachromosomal elements. Plasmids originally found in nature have been modified, shortened and reconstructed to enhance their ability either for general purposes or to suit (fulfil) particular experimental designs, literally for using with different types of host cells.

An ideal plasmid vector (Figure 17.1) for molecular cloning must have minimum amount of DNA, replicate in a relaxed manner rather than in stringent mode to ensure a good yield of DNA, contain at least two selectable markers and have only single recognition site for restriction enzyme. The last specification permits cleavage of the circle at a unique site for ligation to the insert segment. For maximum convenience in selection, the unique restriction site should be within one of the two selectable marker genes. It cannot interrupt sequences that are essential for plasmid maintenance. Vectors that approach such features have been constructed from naturally occurring DNA molecules using both classical genetics and recombinant DNA techniques.

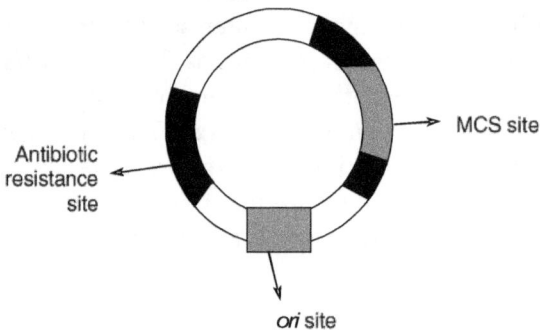

Figure 17.1   Simple structure of a plasmid

Hence a plasmid, (especially for genetic engineering use) is one which contains the following important characteristics.

1. It should be small (<40 kb), supercoiled, covalently closed circular (CCC) DNA and exist as an extrachromosomal element.

2. It should carry a multiple cloning site or polycloning region for insertion of the gene of interest.

3. It should possess at least two selectable markers one of which may be an antibiotic resistance gene.

4. It should be present in large numbers per cell.

5. It must have an origin of replication.

6. It may or may not contain a promoter to express the gene of interest.

Plasmids are classified into various groups based upon various characteristics.

Based upon the ability to take part in conjugation, plasmids are of two types. Conjugative plasmids are those which take part in conjugation. They have *tra* genes which help in conjugation, e.g. F⁺. Plasmids which lack them are called as non-conjugative plasmid, e.g. pBR 322. Plasmids which are used in genetic engineering are of non-conjugative type, so that the gene which is placed does not move from one species to another. But recent reports state that even the non-conjugative type of plasmid can move across species when cohabited with conjugative type of plasmid.

Based upon the number of copies per cell, plasmids are classified into two types.

1. *Stringent plasmids* These plasmids exist in small numbers, i.e., <100 copies/cell. Stringent plasmid is under the control of bacterial genome for replication and

segregation. Generally, conjunctive plasmids are mostly stringent plasmids.

2. *Relaxed plasmids* These plasmids exist in large numbers, i.e., >100 copies/cell. Relaxed plasmid is not under the control of bacterial genome for replication and segregation. Generally, relaxed plasmids are of low molecular weight and most of them are of the non-conjugative type.

The most widely used method to find the copy number of the plasmid is to estimate the amount of enzyme encoded by genes present in the plasmid. For example, β-lactamase activity can be measured if the plasmid specifies ampicillin resistance.

Sometimes plasmids are also classified into compatible groups, based upon plasmid incompatibility. Plasmid incompatibility is the inability of two different plasmids to co-exist in the same cell in the absence of selection pressure. But this method is not widely used.

pBR 322

pBR 322 is one among the most popular and widely used plasmids. It was constructed by using both classical genetic techniques and recombinant DNA methodology. It is one of the most versatile plasmids with characteristics of an ideal plasmid. pBR 322 is 4363 bp in length with Col El as replication start and with two antibiotic resistance genes (ampillicin and tetracycline).

pBR 322 has recognition site for more than 30 restriction enzymes and most of them occur once, thus generating a full-length linear molecule with a cohesive or blunt-end. Nine restriction sites are located in the tetracycline gene, whereas ampillicin has sites for only three restriction enzymes. Thus, cloning in pBR 322 with the aid of any one of those 12 enzymes will result in insertional inactivation of the ampillicin and tetracycline resistance markers. However, cloning of gene in other sites will not permit the easy selection of recombinants because neither of the antibiotic resistance determinates is inactivated. Normally, when pBR 322 is transfected into *E. coli,* K12 cells can yield more than 10^8 transfer or recombinant bacteria. This value depends on various factors hence it cannot be assessed as standard. *E. coli* K12 cells containing pBR 322 grow in media containing either ampillicin or tetracycline or both. When a segment of DNA is inserted into the region of pBR 322 encoding resistance to tetracycline, transformants still grow in ampillicin but not tetracycline. The reverse is applicable when the insert is placed within the ampillicin gene.

In general, selection will be based upon the ampillicin inactivation. The gene of interest is cloned into the ampillicin gene so that the plasmid loses its ability to resist ampillicin antibiotic. The transformed bacteria are plated on soluble starch-rich medium containing tetracycline. After 24 hours the plates are flooded with indicator solution of iodine and penicillin. The bacteria without insert will produce β-lactamase (Ap^r gene) which converts penicillin to penicillonic acid. Penicillonic acid will then bind to iodine and prevent its action on starch. Then the colony of bacteria without insert will be white whereas the bacteria with insert will be blue in colour as it does not produce β-lactamase as the application gene is inactivated by insert.

pBR 322 is the most widely used cloning vehicle. In addition, it has been widely used as a model system for study of prokaryotic transcription and translation.

pUC Plasmids

pUC plasmids stand for plasmids developed at University of California. pUC plasmid was contructed by J. Messing and

J.Vieira. This plasmid contains 40% DNA similar to pBR 322 and has ampillicin resistance gene. It has a short stretch of DNA with a cluster of restriction enzyme recognition sites. This stretch is called as "multiple cloning site" (MCS).

Instead of tetracycline gene, pUC vectors have a *lacZ* gene coding for β-galactosidase enzyme. This gene helps in identification of plasmid pUC with insert and without insert. If we plate these colonies on a medium containing β-galactosidase indicator, colonies with the pUC plasmid will change colour. When bacteria are plated on the medium with X-*gal* and IPTG, the β-galactosidase enzyme will break X-*gal* (colourless) to release galactose and X- (indigo dye), which stains the bacterial colony blue.

On the other hand if we have interrupted the plasmids by placing a gene of interest is the multiple cloning site, the gene becomes inactivated. Hence the bacteria fails to be stained and remains white. Thus, it is a simple matter to pick the colonies with inserts as they are the white ones and bacteria without insert are blue. Notice that in relation to cloning an insert in pBR 322, it is easier to clone in pUC vectors as we have to just look for a clone which is growing on the medium with white in the presence of X-gal.

BACTERIOPHAGES

Bacteriophages are viruses which infect the bacteria. It is one of the virus which is extensively studied, hence various scientists tried to develop a bacteriophage as vector due to its ability to carry a large size of the insert. The DNA of phage is a linear double-stranded DNA of size 48.5 kbp. At the end are short single-stranded, 5′ projection of 12 nucleotides, which are complementary in sequence and by which the DNA adopts a circular structure when it is injected into its host cell. This short stretch of single-stranded complementary sequence is called as 'cos' sequence. Genes on the left of the convential linear map code for head and tail protein of the phage particle whereas the genes on right are concerned with gene regulation and prophage immunity to superinfection. The genes at the centre are concerned with recombination and lysogeny. Much of this central region including these genes is not essential for phage growth and can be deleted or replaced without seriously impairing the infectious growth cycle.

Fred Blatter and his colleagues were the first to develop a bacteriophage as vector. They took out the region in the middle of the phage DNA which codes for proteins needed for lysogeny, but retained the genes needed for lytic infection, but their missing genes can be replaced with foreign DNA. Blatter named these vectors as charon phages. However, for DNA to be packaged into bacteriophage particles, it must be larger than 38 kbp and smaller than 52 kbp in length. Bacteriophages that are available nowadays are excessively modified from that of wild type. Vector bacteriophages are only 33.5 kbp in length, with genes required for a lytic infection kept intact. Only two *Eco* RI sites are present instead of five in the wild type. The central region which is dispensable or replaced with vector can be obtained by digesting the DNA by *Eco* RI. Three fragments generated by the digestion can be purified by agarose gel electrophoresis because of size difference.

To form a recombinant, the two purified arm fragments are mixed with insert fragments, annealed and then ligated together. Sometimes, two inserts can bind to either arm in reverse orientation through cohesive ends and both the arms are attached. Such structures are called as concatemers. These recombinant molecules can be introduced into *E. coli* by transformation and recovered as phage particles after infecting the transformed bacteria with phage λ. The other method is *in vitro* packing of the recombinant molecules in the phage head. Z DNA is packaged *in vitro* by mixing concatemeric DNA, under appropriate conditions, with preformed empty heads, tails and terminase (product of gene E). The head particles are obtained by growing phages which have mutation in (terminase) gene E and in tail genes, whereas tail portion is obtained from phages which have mutation in head genes. When cell-free and DNA-free extracts of the two kinds of infected cells are mixed, intact empty phages are formed.

If recombinant DNA molecules are added into the cell-free and DNA-free extracts, recombinant phages carrying gene of interest are formed. During the packing, head proteins combine to form empty phage. During the packing, head proteins combine to form empty prohead. Then these proheads interact with a complex formed of concatemeric λ DNA and another λ protein, terminase. First, terminase binds to the empty head, then DNA binds to this complex. The DNA is pushed into the head of phage. Then once the terminase encounters the *cos* site in the DNA, it cleaves the DNA at the *cos* site. Then the tail is attached, thus completing the packing.

A large array of λ vectors has been constructed for different purposes. They all share two basic characteristics. The vectors themselves can be propagated as phages in a lytic infection so that vector DNA can be prepared. Carefully positioned restriction endonuclease sites permit cleavage or removal of the central dispensable region and also provide the site for ligation of an insert. As with plasmid vectors, the desired recombinant phages must be selected from among a variety of extraneous products. To distinguish recombinant from nonrecombinant phages, some λ vectors include a DNA segment that contains functional β-galactosidase gene. Most of *lacZ* gene is discarded when the dispensable central gene region is removed by restriction endonuclease cleavage. Consequently, reconstituted vector DNA, which can form β-galactosidase, produces blue plaques on agar plates containing X-gal, whereas recombinant vector

DNA, which does not produce β-galactosidase generates colourless plaques on the same culture medium. Phages containing an insert are thus distinguished by their colour on indicator plates designed to demonstrate the presence or absence of *β*-galactosidase.

M13 Phage

M13 is one of the filamentous bacteriophages of *E. coli*. Its genome is a single-stranded circular DNA molecule about 6408 bp long and is packaged in a tube-like capsid constructed of at least three different phage-encoded proteins (Figure 17.2). M13 only infects strains of enteric bacteria containing F pili on their membrane. The F pilus is required for attachment and adsorption of the phage to the cell. The first step in replication of M13 is converting the single-stranded DNA to a double-stranded circular form of the genome. This is accomplished by synthesis of a strand

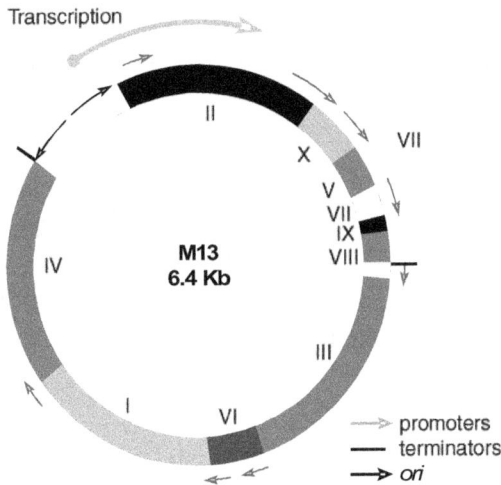

Figure 17.2 M13 phage vector

complementary to that of the infecting DNA. The duplex DNA directs the synthesis of about 300 copies of itself, which then serve as template to produce single-stranded phage DNA for packaging. Replication of phage DNA does not result in host cell lysis, rather, infected cells continue to grow and divide, albeit at a rate slower than uninfected cells, and extrude virus particles. Up to 1000 phage particles may be released into the medium per cell per generation.

The actual molecules used for the construction of recombinants are always the double-stranded replication intermediates. Single-stranded DNAs are not used as vectors because they are not generally cleaved by type-II restriction endonucleases. Like bacteriophages, M13 phage does not carry any unwanted or dispensable DNA. It has only a short stretch of DNA of 507 bp which can be removed without interfering with the viral replication. Hence a *lacZ* gene is inserted in this region for blue–white selection.

M13 host vector systems have several advantages.

1. Very large insert can be cloned, as the size of the phage particle is decided by the size of the viral DNA. Hence there are no insert size restrictions like plasmid. M13 host vectors can carry an insert of up to 95 kbp size.

2. The phage DNA is replicated via double-stranded circular DNA intermediates. These circular DNA can be isolated and transfected as plasmids.

3. M13 phages can provide a single-stranded DNA molecule which is used in DNA sequencing.

COSMID

Cosmids are plasmids with "cos" site. Cosmids are a type of hybrid vectors that replicate like a plasmid but can be packaged *in vitro* into λ phage coats. A typical cosmid has replication functions, unique restriction endonuclease sites and selective markers (*lac*Z, antibiotic resistance contributed by plasmid DNA, combined with a λ DNA segment that includes the joined cohesive ends (cos site) (Figure 17.3).

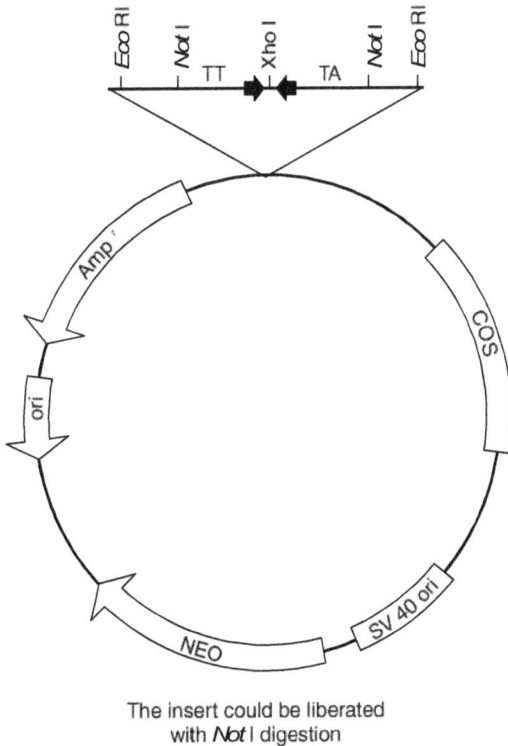

The insert could be liberated
with *Not* I digestion

Figure 17.3 Cosmid vector

Cosmids are used as gene-cloning vectors like plasmids and phages by using *in vitro* packing system. New phage particles are not produced because cosmids do not carry any of the standard λ genes. But *in vitro* packed phage particles can infect a suitable host. The recombinant cosmid DNA is injected and circularizes like phage DNA but replicates as a normal plasmid without the expression of any phage function. This procedure selects for long inserts because the distance from one *cos* site to another must be between 38 and 52 kbp to be packaged in λ head. Thus, a cosmid can carry an insert of size 45 kb (apart, from its own normal size of 5 kb).

Cosmids provide an efficient means of cloning large pieces of foreign DNA. Because of their capacity for large fragments of DNA, cosmids are particularly attractive vectors for constructing libraries of eukaryotic genome fragments. When transfecting the cosmid, fake dummy phage particles are mixed as this adding enhances the transfection efficiency. Between 1×10^4 and 5×10^5 colonies of transformed cells are obtained per µg of insert DNA used. Because of the presence of plasmid elements, the recombinant transformants can be selected by using blue-white selection.

PHASMIDS

Phasmids are truly plasmids with phage genes. They are linear duplex DNAs whose ends are λ segments that contain all the genes required for a lytic infection and whose middle is linearized. Both the λ and the plasmid replication functions are intact. Normally, plasmid vectors carry a λ attachment site. Once inside the *E. coli* cell, the plasmid can replicate like a phage and form plaques in the normal way. However, if the vector contains the gene that encodes λ repressor, then the phasmid replicates as a plasmid rather than as a phage. Depending upon the functioning or non-functioning of cI–Protein (coded by λ–repressor), the phasmid can replicate as plasmid (cI–Protein inactive) or phage when cI–protein is active. The activity of cI–protein can be inactive by growing the *E. coli* culture at 40°C.

Plasmids may be used in variety of ways. For example, DNA may be cloned in the plasmid vector in a conventional way and then the recombinant plasmid can be lifted onto the phage. Phasmids are easy to store, they have an effectively infinite shelf-life and screening phages by molecular hybridization gives cleaner results than screening bacterial colonies.

SHUTTLE VECTOR

Shuttle vector is that which can multiply in the two different species. Shuttle reactors are designed to replicate in the cells of two species, as they contain two origins of replication, one appropriate for each species as well as genes that are required for replication and not supplied by the host cell (Figure 17.4). Shuttle vector has advantages of replicating in diverse hosts. For example cloning and isolating DNA segments for structural analysis is most conveniently achieved with *E. coli* host vector. However functional analysis of cloned insert must usually be carried out in species of origin. If a researcher requires to study the above problems then it is best to clone the DNA in a vector which can be shuttled back and forth between each and every alternate cells and replicates in both. Shuttle vector contains two origins of replication, one is specific for host and the other is specific for another host (e.g. yeast *ori* site). The presence of two replication origins sometimes poses special problems, one portion of replication origin of one species is totally unrelated to another and interferes with the replication of the other host. Hence in a shuttle vector various types of replication origins are to be inserted and checked before experimenting.

Nowadays shuttle vectors which function in a given set of species or host are commercially available. For example, one type of shuttle vector is used to clone the gene in *E. coli* and yeast and another type of shuttle vector is used for *E. coli* and animal cells, e.g. SV40 plasmid vector.

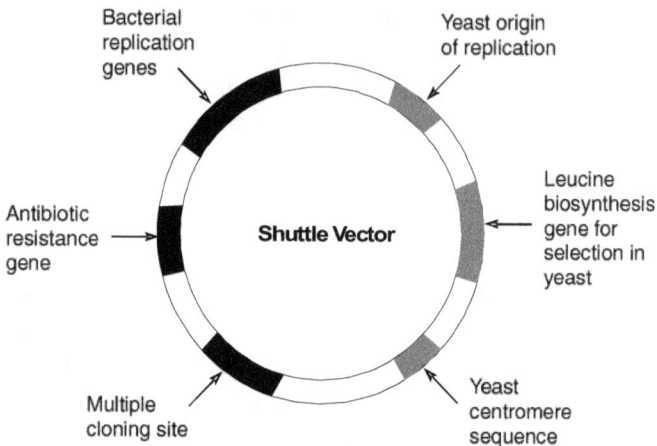

Figure 17.4   Shuttle vector

YEAST VECTORS

Yeast Cloning Plasmid Vectors

The analysis of eukaryotic DNA sequence has been facilitated by cloning the genes in prokaryotes. But some functions such as glycosylation, mitosis, meiosis, etc. are absent in prokaryotes. When genes functionally related to such a function are to be analysed, those genes have to be cloned in a eukaryotic system. Yeast system offers the best system to study, as it is small, and easy to grow and manipulate. In addition, the biochemistry and regulation of yeast is very much like that of higher eukaryotes. Thus yeast can be a very good surrogate host like *E. coli,* for studying the structure and functions of eukaryotic gene products. Various yeast vectors have been designed, once the ability and utility of yeast is confirmed. All of them have three features in common.

1. All of them contain unique target sites for a number of RE.

2. All of them can replicate in *E. coli* often at high copy number.

3. All of them employ markers that can be used to select recombinant yeasts, e.g. Hi53, leo2, trp1 and ura3.

Yeast Episomal Plasmid (YEP)

These are naturally available plasmids of size 6.3 kb. This plasmid is also called as 2 μm plasmid. YEP contains an origin of replication and three genes, REP3 (*cis*-acting elements) REP2 and REP1 (Figure 17.5). 50–100 copies of plasmids exist per cell, which represent 2–4% of total yeast genome. About 50% of YEP is essential for replication and maintenance and the remaining is dispensable. These plasmids have no markers and both the halves are useful. Hence Beggs in 1978 constructed a shuttle vector by using the essential portion of 2 μM plasmid and pBR 322. The marker gene encodes the enzyme required for histidine biosynthesis. Hence the recombinant transformants were selected by growing the yeast mutant for histidine biosynthesis on a medium lacking histidine.

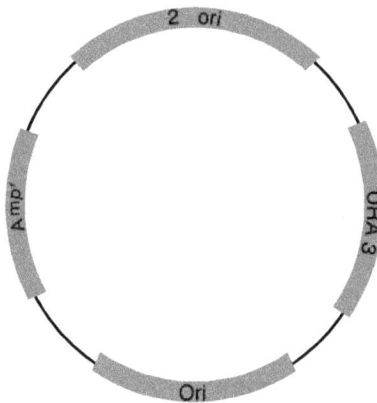

Figure 17.5   Yeast episomal plasmid

Yeast Replicating Plasmid (YRP)

YRP incorporates a yeast genomic replication origin into a circular duplex DNA (Figure 17.6). It contains a 100 bp sequence called autonomously replicating sequence (ARS). Any circular duplex that contains the ARS can be used directly for cloning in the yeast. Although plasmids containing ARS transform yeast very efficiently, the resulting transformants are exceedingly unstable. For unknown reasons, YRP plasmid tends to remain associated with the mother cell and is not efficiently distributed to the daughter cell. Even in the mother cells, the molecules are stably maintained with yeast cell so long as they are the sole source of a required gene product (e.g. *trp* gene in trp⁻ host). Once the host is not subjected to selective pressure, the plasmid is lost. (*trp⁻* host containing trp gene is grown on tryptone-rich medium).

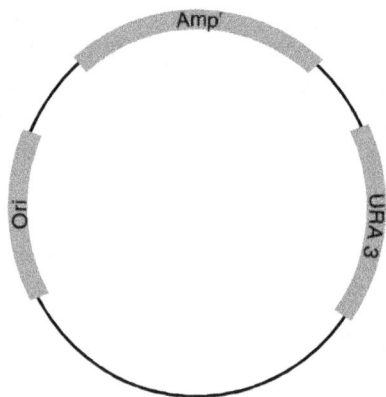

Figure 17.6   Yeast replicating plasmid

Yeast Centromere Plasmid (YCP)

YCP contains a short segment of size 1.6 kb, which functions as centromere. This centromere is required for segregation of the YCP during the yeast division. YCP behaves like a small functional chromosome; hence it is also called as mini-chromosome vector. YCP contains Amp resistance gene and *leuZ* gene, so that they can be selected both in *E. coli* and in yeast. Apart from these, YCP also contains a single copy of ARS sequence. Functionally YCPs exhibit the characteristics of chromosomes in yeast cells. First, they are mitotically stable in the absence of selective pressure. Second, they segregate during meiosis in a Mendelian manner. Finally, they are found at low copy number in the host cell. These vectors are used when the gene product is deleterious to the cell or when complementation studies are required. Low copy number makes recovery of plasmid from yeast very easy.

Yeast Artificial Chromosomes (YACs)

All three autonomous plasmid vectors are maintained in yeast as circular DNA molecules. Thus none of these vectors resembles the normal yeast which has a linear structure. Yeast artificial chromosome is a linear chromosome and contains a telomerase along with *leu*+ gene (Figure 17.7). Because of the presence of telomerase, there exist 1–2 copies of YAC in each cell. These vectors can carry an insert of 40 kb length. Apart from that they can amplify large DNA molecules in a simple genetic background.

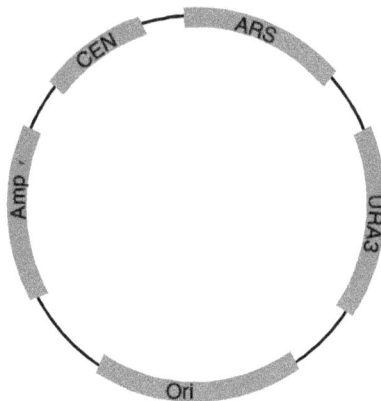

Figure 17.7 Yeast artificial plasmid

ANIMAL VECTORS

Baculovirus Vector

Baculovirus infects insects. This virus is rod-shaped with a large double-stranded genome. During normal infections, baculovirus produces nuclear inclusion bodies which consist of virus particles embedded in a protein matrix. This protein matrix is encoded with the virus and is called polyhedrin and this polyhedrin accounts for 70% of total protein encoded by the virus as the transcription of the polyhedrin is driven by extremely active promoters.

Genetic manipulation of the viral DNA is not possible as it has a very large DNA with many restriction sites for a single enzyme. Hence, the gene of interest is cloned into the small recombination transfer vector and co-transfected into insect cell lines along with the wild type of virus in the cell. Homologous recombination takes place between the polyhedrin gene and our gene of interest. Thus, our gene of interest will be transferred from the vector plasmid into the wild type of virus, polyhedrin gene will be transferred from the virus on the plasmid. This is something like displacement reaction. This displacement of gene will not effect the replication of virus, as polyhedrin gene is not required for replication. The recombination virus replicates in the cells and generates characteristic plaques (without inclusion bodies). Normally the virus is cultured in the insect cell line of *Spodoptera frugiperda*. The foreign gene is expressed during the infection and very high yields of protein can be achieved by the time the cell lyses.

SV40

SV40 is a spherical virus with double-stranded circular DNA of size 5.2 kb. The viral protein contains three viral coded proteins.

VP1 is the major protein present in the capsid with a size of 47000 kDa. Two more proteins VP2 and VP3 are also present. The DNA of virus is associated with the four histones (H4,H2A,H2B and H3) proteins. The viral DNA can be segmented into five precise segments coding for five different proteins small T, large T, VP1, VP2 and VP3. VP1 coding region overlaps VP2 and VP3 in a different translation reading frame. SV40 virus infects monkey kidney cell lines. The virus travels to the nucleus and gets uncoated. Then both the T-genes located near the origin are translated in the clockwise direction. The large T protein is important for virus DNA replication and starts after the translation of large T-protein. Replication starts at the origin and is bi-directional. It terminates when two replication forks meet. About 10^5 molecules of duplex DNA are synthesized per cell. Along with DNA replication, VP1, VP2 and VP3 proteins are synthesized. Then packing of DNA occurs to form new virions, which are released by the lysis of cell. The entire process can also be initiated by transfection with naked SV40 DNA. SV40 vectors are constructed similar to phage vectors. Portions of the viral genome are removed and replaced by other DNA segments. There are three types of SV40 vehicles each of which have a distinct advantage or disadvantage among themselves.

SV40 transducing vectors   These vectors are capable of replicating and packing into virion particles. Transducing vectors contain a segment of 300 bp which functions as the origin of replication and provides the transcriptional regulatory signals for the synthesis of mRNAs. This type of vector takes an insert of size 3.9 to 4.5 kb. These plasmids do not have the genes that code for VP1, VP2 and VP3. As no DNA can be added to SV40 DNA without removing any DNA from the genome, to add the insert, the genome DNA that is not required is removed. The functions of the DNA that are lost by these deletions are supplied by using a helper virus or by inserting the SV40 deleted genes into the host DNA so that the proteins are supplied in *trans* mode. Normally the recombinant SV40 vectors (usually consist of DNA of interest and replication sequence and gene for coding VP1, VP2 and VP3) are transformed into the COS cell line. COS cell line is a kidney cell line of the African green monkey kidney. It has the T-protein gene incorporated in the genome. So when the vector is transfected into these cells, virion particles are yielded with the help of helper virus.

SV40 plasmid vector　These vectors multiply in the monkey cell line but are not packed as the virions. These plasmids/ vectors usually contain origin of replication sequences and larger T-protein gene but do not contain VP1, VP2 and VP3 genes. They are shuttle vectors, and have the ability to multiply both in *E. coli* and monkey cell line. Normally the recombinant is multiplied in *E. coli* cells to high copy number and then transferred into the cell line. These cells are stable in bacterial cell and are efficiently transferred from parent cells to daughter cells. However, the plasmid vectors are unstable in most animal cells and cannot be maintained indefinite.

SV40 passive transforming vectors　These vectors neither replicate nor produce virions, but simply integrate the DNA segments into the cellular DNA. These transformed cells replicate the new DNA as an integral part of their own genomes. These plasmids are also shuttle vectors and include selective markers like herpesvirus, thymidine kinase or neo genes. Apart from the selective markers, they include transcriptional regulator signals and polyadenylation sites.

Bovine Papilloma Virus Vector

Bovine papilloma virus (BPV) causes warts (uncontrolled epithelial proliferation) in cattle. It is a member of a group of viruses which induces warts and papillomas in a range of mammals. BPV normally infects terminally differentiated squamous epithelial cells.

BPV has a capsid protein surrounding a circular double-stranded DNA of size 79 kb. 69% of this genome is important for viral function, whereas 31% of the genome can be replaced by the insert. The recombinant BPV is constructed by ligating the insert and BPV vector (69%) onto the pBR 322 plasmid, thus generating the shuttle vector containing plasmid *ori* site and virus replication sequences. These shuttle vectors are multiplied in *E.coli* cells first and then they are transformed into mouse cell line. It has been observed that if these sequences are removed prior to transfection, the vector exists at high copy number i.e., 200 copies per cell. When transfected with pBR 322 sequences, it exists at low copy number, i.e., less than 10 copies/cell.

The major advantage of BPV is the generation of permanent cell line. As the infected cells are not killed, a stable plasmid number is found even when the insert is of large size. The selection of transformants is very easy as they form a pile of cells on the transferred monolayer of cells called "Focus". The transformed cells are then selected by the presence of marker gene which is mostly the neomycin phosphotransferase gene coding for resistance against G418.

PLANT VECTORS

Like bacteriophages, even plant viruses can transfer recombinant genes into plants. Plant viruses are attractive as vectors for several reasons.

1. Plant viruses adsorb to and infect cells of intact plants.

2. Relatively large amounts of virus can be produced from infected plants, leading to the prospect of large amount of foreign protein being expressed from recombinant viruses.

3. Some virus infections are systematic. They are spread throughout the whole plant. In some cases intact viruses are transported through the vascular system of plant.

Even with interesting characteristics, at present, plant viruses have not been developed to the stage where they are widely used as vectors, but progress has been made only with the two groups of plant viruses, cauliflower mosaic viruses (CaMV) and Gemini viruses, both DNA viruses.

Cauliflower mosaic virus (CaMV)　CaMV is a circular, double-stranded DNA genome of about 8 kb. It causes systemic infection in small groups of plant species. It contains three discontinuous duplex DNA, (two in one strand, one in the other). These are regions of sequence overlap. DNA itself is infected and can commence infection by inoculating on the surface of the host plant. CaMV genome contains eight closely packed reading frames. There are only two small intragenic regions which can be discarded and replaced by the insert. If any other region is removed or when a large-sized insert is used, no infection takes place. Thus, only a small size of insert of few hundred bp can be used as insert. This requirement of a small-sized insert imposes limitations on the use of this vector for gene transfer. Even the use of helper virus coding for the lost genes when a large insert is placed will not restore their functions. The reasons are not clear.

T-DNA PLASMID

Plants do not have plasmids indigenous to them as yeasts do. Hence, bacterial plasmids, which can transfer DNA to plants were developed. The best-developed plasmid-based vector to transfer DNA into the plant is Ti-plasmid (Figure 17.8). Ti-plasmids occur naturally in gram-negative bacteria called *Agrobacterium tumefaciens.* Ti-plasmid is 2.5 kb in size with unique regions called A,B,C and D. Regions B and C are involved in plasmid replication. D region is involved in the transfer of DNA from the plasmid into plant. This region of D is 40 kb in length and is referred to as the virulence region.

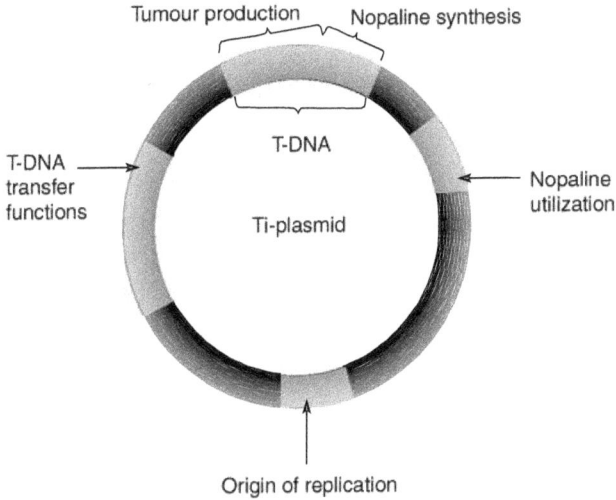

Figure 17.8 Ti-plasmid

vir genes always transfer a fixed region from the plasmid to the plant cell. Such region is called as T-DNA or transfer DNA. T-DNA is 13–25 kb in length and codes for tumour-inducing proteins. T-DNA integrates into host genome at random sites without any specificity. These cells are transformed and produce a tumour called a "crown gall". T-DNA codes for 3 proteins, 2 of them are responsible for the synthesis of auxin and cytokinin. The other protein directs the synthesis of unusual amino acids or sugar derivatives called as opines. Opines are not synthesized by untransformed cells.

Ti-plasmids are classified into two types based upon whether they produce octopine or nepaline. Ti-plasmid thus provides bacteria with two important resources—a source of metabolite and the means to use that metabolite as a source of energy.

Ti-plasmids are remarkable because they stand as examples for the insertion of prokaryotic gene into a eukaryotic genome. The plasmid looks like a natural chimeric as it contains two sets of genes, one active in bacteria and the other in plant. The genes in the T-DNA segment are associated with transcription control signals that operate in plants while those in the remainder of the plasmid are under the control of bacterial promoters.

T-DNA transfer is remarkable from the point of view of interkingdom gene transfer. Although T-DNA is transferred, it has no role in the transfer mechanism. T-DNA region in both Ti and Ri plasmids are flanked by almost perfect 25 base pair direct repeat sequences. Especially the right hand 25 bp sequence is compulsorily required for T-DNA transfer as they function in a *cis*-acting manner. Any DNA sequence could be transferred to plant cells as long as it is flanked by the 25 bp repeat sequence in correct order by *vir* genes.

Virulence region consists of six genes called *vir A, B, C, D, E* and *G*. *vir* genes are responsible for T-DNA transfer from the bacteria. Out of these, *vir A, B, D* and *G* are compulsory for the T-DNA transfer. The *vir A* gene codes for a protein *vir A* which is located on the inner membrane of the bacteria and is a chemoreceptor which senses the presence of acetosyringone. Acetosyringone is present in plant exudates and is synthesized when a plant gets a wound. Once acetosyringone binds to *vir A* product, it phosphorylates *vir G*, which then stimulates transcription by binding to operate sequences in the promoters of the inducible *vir*-genes. Once *vir D* protein is made, it binds to 25 bp repeat sequences and nicks between third and fourth base pair. The single T-DNA strand is released by replacement synthesis of DNA in a 5´–3´ direction, initiating from the nick in the right-hand 25 bp repeat sequence. *vir E* protein binds to the single-stranded DNA and stabilizes the structures during the transfer. *vir B* protein helps in directing T-DNA transfer extracellularly. In essence, any *Agrobacterium* strain containing a set of functional *vir* genes is capable of transferring any DNA sequence into a plant cell, provided that the DNA is placed between two correctly oriented T-DNA border 25 bp repeat sequences.

REVIEW QUESTIONS

1. Explain about pBR 322.
2. How is recombinant pUC19 selected?
3. Give an account of bacteriophages used in genetic engineering.
4. Write in detail about cosmids and phasmids.
5. Explain about vectors which multiply in yeast.
6. Give an account of about animal vectors used in genetic engineering.
7. Write short notes on the following.

      i.   Characters of ideal plasmid
      ii. Stringent plasmid
      iii. Relaxed plasmid
      iv. M13 phage
      v.   Shuttle vector
      vi. Yeast episomal plasmid
      vii. Yeast replicating plasmid
      viii. Yeast centromere plasmid
      ix. Yeast artificial plasmid
      x.   SV40
      xi. T-DNA plasmid

18
BLOTTING TECHNIQUES

Blotting is the technique in which nucleic acids or proteins are immobilized onto a solid support generally nylon or nitrocellulose membranes. Blotting of nucleic acid is the central technique for hybridization studies. Nucleic acid labelling and hybridization on membranes have formed the basis for a range of experimental techniques involving understanding of gene expression, organization, etc.

SOUTHERN BLOTTING

The original method was first developed by E. M. Southern in 1975 and later was modified by various people for various applications.

The process of Southern blotting starts after the agarose gel electrophoresis of restriction digest or gDNA. The gel is first washed with buffer to remove the broken or fragmented residues of agarose formed during banding and the accumulated contaminants. Then the gel is placed in acid solution (0.25 M HCl) for about 5–10 min. During this step, the DNA undergoes depurination. After this step the gel is placed in 0.25 M NaOH alkali solution. This step ensures that the DNA is shorter in length (<500 bp) and is in single-stranded form. DNA fragments large in size (>10 kb) require a longer time and move in zig-zag form when compared to shorter DNA fragments. Finally, the gel is extensively washed with buffer to remove the traces of HCl and NaOH. Then the gel is placed on the filter paper with wigs dipped in a reservoir containing transfer buffer placed in transfer buffer reservoir tank which is placed on the glass plate (Figure 18.1).

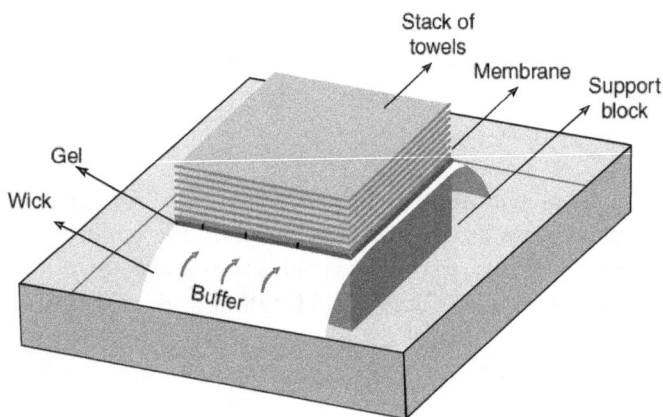

Figure 18.1   Diagram showing blotting set-up

Nitrocellulose or nylon filter paper is placed on the gel above which a stack of blotting papers soaked in transfer buffer is placed and gently pressed using glass plate or rod to remove the air that is trapped between the gel and membrane. Then a stack of unsoaked blotting papers is placed above them and the glass plate with a weight of 500 g is placed above it. By capillary action the transfer buffer is transferred from the reservoir to the blotting papers. During this process the DNA from the gel is transferred onto the nitrocellulose paper and immobilized there due to the negative charge of the DNA and the positive charge of the membrane.

This set-up is allowed to stand like that for 12–18 hours. After this time period the nitrocellulose or nylon membrane is taken out from the set-up carefully. As the binding of the DNA to the nitrocellulose or nylon membrane is weak, it has to be fixed more strongly, so that the DNA is not washed off during the washing of the membranes. There are two widely used methods for this procedure.

After the transfer, the membrane is exposed to ultraviolet rays, so that a bond is formed between thymine residues present in the DNA and the positively charged amino groups on the surface of the nylon membrane. This method is called as UV-cross linking method. The other method is baking of the nitrocellulose or nylon membrane at 80°C in a vacuum oven. The use of vacuum oven is a must due to the flammable nature of nitrocellulose membrane.

At present, various advances have been made in the technique of Southern blotting, the first and foremost being the use of electricity instead of capillary action as a force for transferring the DNA from the gel onto the membrane. The second and most significant is the use of positively charged nylon membrane, which reduces the need for excessive depurination. These two alternative methods have significantly reduced the time and damage to DNA during the process.

NORTHERN BLOTTING

It is same as Southern blotting but with minor differences. In this method, instead of DNA, RNA is transferred onto the membrane from the gel. The other difference is the absence of the depurination step (i.e., 0.25 M HCl treatment), as RNA itself is small enough to be transferred easily.

WESTERN BLOTTING

In western blotting, proteins are transferred from the polyacrylamide gel (PAGE) or sodium dodecyl sulphate-PAGE onto the nitrocellulose membrane or nylon membrane. In this method there is no requirement for pre-treatment as the proteins are small and in most cases they are not linear. The second difference is that there is no need for cross linking of the proteins to the membrane. After the transfer of proteins from the gel onto the membrane, it is incubated with or in a solution containing antibodies against protein of interest. Non-bound antibodies are washed off the membrane and the presence of the initial antibody is detected by placing the membrane in a solution containing a secondary antibody. These secondary antibodies react with immunoglobulins or primary antibodies. This secondary antibody is conjugated to either a radioactive isotope or an enzyme that produces visible colour to analyse the protein expression and regulation.

HYBRIDIZATION

Hybridization is the technique in which the nucleic acid or protein immobilized on the membrane is challenged with a probe or antibody known (Figure 18.2). Hybridization is widely used to confirm the presence or absence of the DNA/RNA/protein in the unknown sample. Hybridization depends on the function of the labelled base pair between the probe and the target sequence. After the blotting technique is completed, the membrane is placed in a solution of labelled (radioactive or non-radioactive) single-stranded DNA or RNA solution. This DNA or RNA contains sequences complementary to DNA or RNA present on the membrane. This labelled nucleic acid used to detect or locate DNA is called as probe. Conditions are chosen such that labelled DNA or RNA bind or hybridize with nucleic acid present on the membrane.

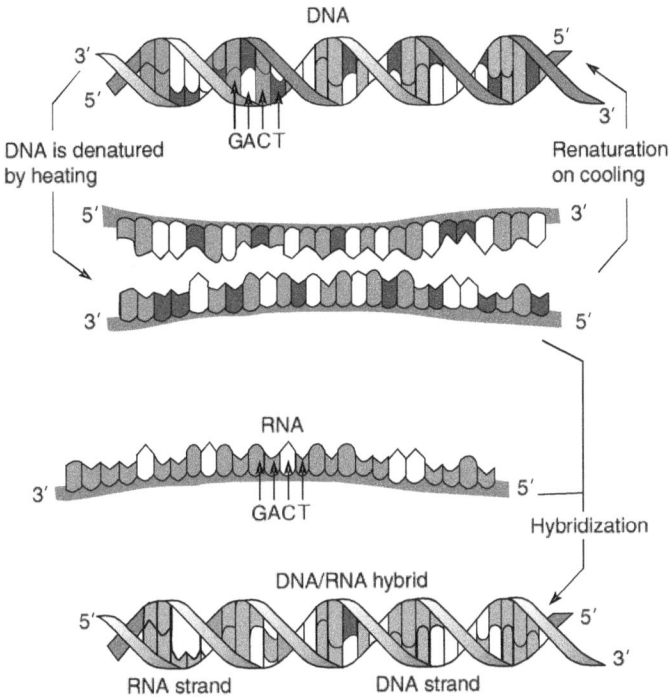

Figure 18.2 Nucleic acid hybridization

If the sequence of nucleic acid in the probe is complementary to nucleotide sequence on the membrane then base pairing or hybridization will occur. The nucleic acid present on the membrane is single-stranded and is bound on the membrane by using negative charge of the phosphate and thymine molecules. Thus when it finds a complementary strand, it forms or develop hydrogen bonds or converts into hybrid DNA (i.e., double-stranded DNA in which the two strands come from different DNA molecules). Conditions are chosen or maintained such that there is a maximum chance of specific hybridization and minimum of non-specific hybridization.

After the hybridization, the membrane is washed to remove the unbounded probes, while bound probes remain attached. The regions of hybridization are detected by autoradiography method (if the probe is radioactively labelled) or by biotin streptavidin method (if the probe is labelled by a non-radioactive). The specificity with which a particular target sequence is detected by hybridization to a probe is called as stringency. At high stringency, hybridization occurs when the sequence on the membrane is completely complementary to the sequence of the probe. In practice, the hybridization is carried out at low stringency level, and the membrane is washed many times such that only perfect (>80%) matches are left over.

The principle of hybridization is more or less same for all, but to differentiate and make it conceptually undoubtful, various names are given. If the probes contain or are made of DNA, they are called as DNA probes and if they are made of RNA they are called as RNA probes. If the membrane containing immobilized DNA is challenged with DNA probe, it is called as DNA hybridization. If the membrane containing immobilized DNA is challenged with RNA probe, then it is called as DNA : RNA hybridization. Similarly, if the membrane containing immobilized RNA is challenged with RNA probe then it is called as RNA hybridization and if challenged with DNA then it is called as RNA : DNA hybridization.

Application of Blotting and Hybridization Techniques

1. Southern blotting technique is widely used to find specific nucleic acid sequence present in different animals including man. For example if we want to know whether there is a gene like insulin in sea anemone, then DNA of sea anemone is mobilized on membrane and blotted by using insulin probes against it.

2. Northern blotting technique is widely used to find gene expression and regulation of specific genes. For example if we find human insulin-like gene in oyster, then by isolating and immobilizing RNA and blotting it with insulin probe we call tell whether the gene is expressing or not.

3. By using blotting technique we can identify infectious agents present in the sample.

4. We can identify inherited disease.

5. It can be applied to mapping restriction sites in single copy gene.

Drawbacks

1. The process is a complex, cumbersome and time-consuming one.

2. It requires electrophoretic separation.

3. Only one gene or RNA can be analysed at a time.

4. Gives information about presence of DNA, RNA or proteins but does not give information about regulation and gene interaction.

DOT BLOTTING TECHNIQUES

The drawbacks of blotting techniques have lead to the development of dot blotting technique which is more advanced, less time-consuming, accurate and applicable to a wide variety of gene/source simultaneously. The dot or slot blotting technique is the most widely used of all techniques for analysing. None of the blot methods require electrophoresis prior to blotting and hybridization. Hybridization of cloned DNA without electrophoretic separation is called as dot blotting.

PLAQUE/COLONY BLOTTING TECHNIQUE

This method was first developed by Granstiens and Hogness (1975). This method is used to identify which colony of bacteria contains the DNA of interest among thousands. In this procedure, the bacterial colonies to be screened are transferred onto nitrocellulose or nylon membrane by using replica plating. Due to the negative charge of the cell surface, some cells bind to the nitrocellulose membrane. Then the membrane is placed in a solution of 0.5 N NaOH to break the cell surface, convert dsDNA to ssDNA and to bind DNA to the membrane. Later, the membrane is transferred to a solution containing protease solution after neutralizing with neutralization solution. The DNA is fixed tightly to membrane by either UV cross-linking or oven baking.

This membrane is used for hybridization with a probe and analysed by using autoradiography or biotin method for positive hybridization. A colony whose DNA print (as replica plating provides a replica print master plate colony on the membrane) gives a positive hybridization can be picked from the master plate. Plaque-blotting is similar to colony blotting, the only difference is that instead of bacterial colony, a plaque is transferred onto the membrane. Benton and Davis developed this method in 1977. The greatest advantage of this method is that several identical DNA prints can be easily made from a single master plate containing bacteria/plaques which are to be made.

DOT BLOT ASSAY

This method is widely used to hybridize DNA from a single cell type against a wide variety of probes, for example, for a viral infection which cannot be identified by normal conventional methods or if we want to know what all genes are expressed in a single cell type (e.g. brain cell). Cell type or cells that are to be screened are placed on the membrane as 'dot' in the order of rows and columns. Then the cells are denatured by using enzymes or detergents (SDS) and DNA is fixed by using UV- cross link or oven baking. This membrane is then used for hybridization by using probes (which are specific to a gene).

REVIEW QUESTIONS

1. Describe in detail about Southern blotting.
2. Enumerate the difference between Southern, western and northern blotting.
3. Write a detailed account of hybridization.
4. Describe the applications and drawbacks of hybridization.
5. Explain the difference between dot blot and Southern blotting.
6. Write a short note on dot blot assay.

19

GENERATION OF CLONES

GENERATION OF INSERT

Recombinants are usually constructed using a large and complex mixture of potential inserts or genes of interest, thereby generating an extensive collection of clones. The desired recombinant is finally obtained by specific selection and screening procedure. Both procedures become much easier if the initial mixture is partly enriched for the segment of interest because fewer recombination clones have to be tested to find the desired end.

There are four sources of inserts for cloning.

1. Genomic DNA fragmentation by RE which is widely called as shot gun method.

2. gDNA fragmentation by mechanical shearing, widely called as fragmentation.

3. Synthetic DNA preparation by combination of chemical and enzymatic methods.

4. Preparation of DNA segments by enzymatic copying of RNA template.

Fragmentation

The long, thin threads which consist of duplex DNA molecules are sufficiently rigid to be broken by shear forces solution. Intense sonification with ultrasound can reduce the length to about 300 bp. More controlled shearing can be achieved by high-speed stirring in a blender. Typically high molecular weight DNA can be broken down into fragments of about 8 kb by stirring at 1500 rev/min for 30 min. Breakage occurs essentially at random with respect to DNA sequence. The only problem with this method is that the results or the length of fragment is not reproducible.

Shot Gun Method

The direct method of obtaining an insert is to digest the total genomic DNA with restriction enzyme. This procedure has the advantage of reproducibility. The same set of fragments is produced each time a specific enzyme cleaves a particular DNA. If the endonuclease yields cohesive ends that match those on the vector, cloning is carried out directly or after enrichment of the population of fragments for the desired insert. When the digested genome is large, few well-separated DNA fragments are obtained. Rather, a continuous smear of DNA fragments of all possible sizes appear. For example, if human genomic DNA (3×10^9 bp) is subjected to RE digestion which has a six base pair recognition site, then the RE will recognize a recognition site for every 4096 (4^6) bp on average and can yield about 7×10^5 unique fragments. The resulting sets of fragments yield a continuous spread of DNA molecules of all possible sizes. Electrophoresis and/or high performance reverse phase liquid chromatography are widely used to resolve the fragments based on size. However neither method yields pure DNA fragments from complex mixtures. Preparations are usually contaminated with many fragments of similar size or elution properties. Nevertheless, even minor enrichment for the desired fragment can significantly decrease the problem associated with finding the desired DNA segment in a large collection of clones.

cDNA Synthesis

Methods of analysing RNA are laborious and critical when compared to methods of DNA analysis. Hence it is best to convert the RNA into DNA molecule. DNA molecule synthesized enzymatically by copying an RNA template are called cDNAs or complementary DNA. cDNA can be used as DNA insert in gene cloning experiments.

The first step in cDNA synthesis is isolation of the mRNA from the tissue where they are produced in high quantities. These mRNA are placed in tubes containing reverse transcriptase enzyme, a small primer or oligo(dT), and dNTPs. The oligo(dT) binds to the mRNA at its 3'-OH end, as mRNA (especially eukaryotic) contains a poly(A) tail at its 3'-OH end. The oligo(dT) acts as a primer and provides 3'-OH end for the reverse transcriptase enzyme. Reverse transcriptase enzyme adds dNTPs to new complementary DNA by taking RNA molecule as template strand. In the second step, using RNase H or alkali treatment, RNA molecule is degraded. Then the newly synthesized cDNA are separated and added to a fresh tube containing DNA polymerase and dNTPs. No primer is required at this step because the variable length of hairpin loop is formed transiently at the 3'-OH end of the first strand, which serves as a primer for the second strand DNA synthesis. After the second strand is synthesized, the hairpin structure is removed by using S1 nuclease enzyme.

This cDNA, which is double-stranded, is used as insert in most of the experiments. cDNAs have their own special uses, which is derived from the fact that they lack the intron sequence that are usually present in the corresponding genomic DNA. cDNA clones find application or use when bacterial expression of the foreign DNA is necessary as a prerequisite for detecting either the clone or the polypeptide product.

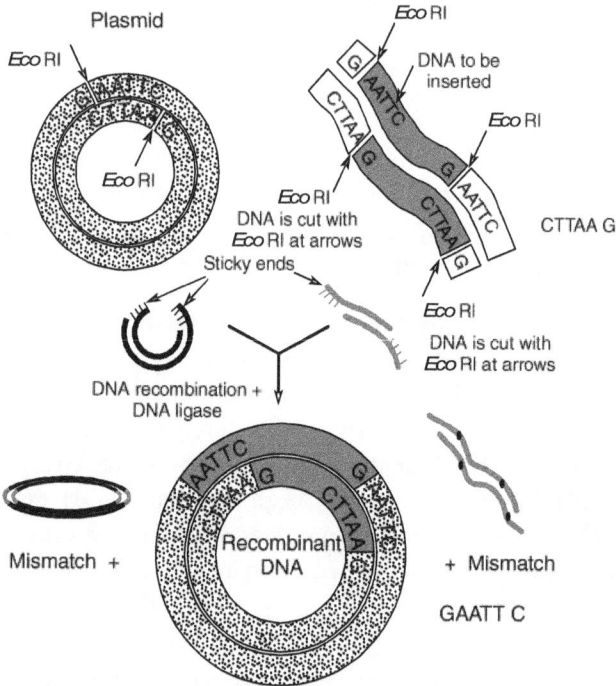

Figure 19.1   Various products formed during ligation reaction between insert and vector

The second step in gene cloning is the joining together of the vector molecule and the DNA to be cloned (Figure 19.1). This process is referred to as ligation. The enzyme which catalyses the reaction is DNA ligase. The optimum temperature for ligation of nicked DNA is 37°C, but at this temperature the hydrogen bonding formed between the sticky ends is unstable. Hence, the ligation reaction is carried out at 4°C. Even though the rate of reaction is slow and takes a long time to complete, this temperature is preferred, as the success rate is high. The ligation reaction of vector and inserts generate 3 types of products 1) the insert gets circularized without ligating to vector 2) The vector recircularizes without insert and 3) the vector–insert gets ligated. The last combination is important. To favour the formation of recombinants, various strategies have been devised. Firstly, performing the reaction at a high DNA concentration can increase the population of recombinants, i.e., the ratio of vector and insert is kept at 1 : 2 so that there is maximum chance of vector and insert collision.

Secondly, by using alkaline phosphatase, the 5′-end of the plasmid DNA is removed, thus the vector cannot religate itself. But the insert can supply 5′-PO_4 to the 3′-OH of vector. Thus a phosphodiester bond will form only when the insert and vectors get circularized, but one nick will exist as the 3′-OH of insert cannot ligate to vector as it does not have a 5′-end. This nick is repaired by host cellular repair mechanism after recombinant molecule enters into the cell. Generation of recombinant molecules or vector–insert is very high when the vector and insert have compatible cohesive ends, i.e., when vector and insert are subjected to same restriction enzyme, e.g. *Eco* RI, which generates cohesive ends of same complementary ends. The cohesive end of the complementary sequence can form hydrogen bonds and can hold vector–insert DNA together temporarily, so that DNA ligase can form the phosphodiester bond. The efficiency of recombinant molecule formation is a bit less when both the insert and vector or one of them has a blunt end. Inserting DNA segments into vectors is always more efficient if the vector and insert have matching cohesive ends. There are two methods to convert the blunt ended DNA fragments into cohesive ends.

Linkers and adapters   Linkers are short stretches of double-stranded DNA of length 8–14 bp and have recognition site for 3–8 RE. These linkers are ligated to blunt-end DNA by ligase. Because of the high concentration of these small molecules present in the reaction, the ligation is every efficient when compared with blunt-end ligation of large molecules. The cohesive ends are generated by digesting the DNA with appropriate RE that generates cohesive ends by cleaving in the linkers (Figure 19.2). The problem with linker is that the sites for the enzyme used to generate cohesive ends may be present in the target DNA fragment. This drawback limits the use of linkers for cloning.

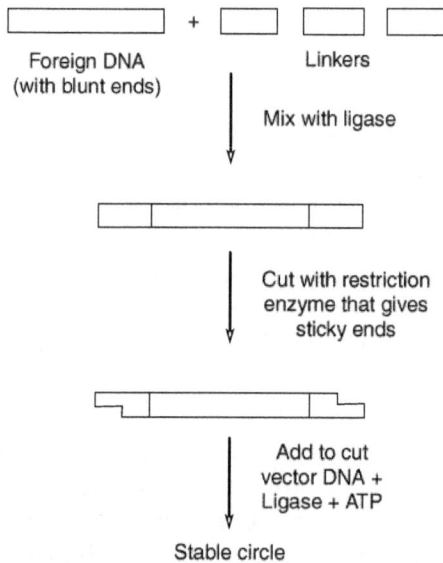

Figure 19.2 Steps involved in converting blunt-end insert into cohesive end

Adapters are linkers with cohesive ends or a linker digested with RE, before ligation. The most widely used definition is cut linkers also called as adapters. They are not perfectly double-stranded nor single-stranded. By adding adaptors to the ends of a DNA, sequences that are blunt can be converted into cohesive ends.

Homopolymer or T/A tailing

This method uses the ability of annealing of complementary strands or sequences. Suppose a vector has an oligo(dA) sequence at the 3'-OH end and the insert has an oligo(dT) sequence at its 3'-OH end. Then when both the molecules are mixed, the molecules are held by hydrogen bond or can anneal until the ligase joins them by phosphodiester bond. The important component in this method is terminal deoxynucleotidyl transferase. This enzyme adds nucleotides at the 3'-OH end of DNA without any complementary sequence. It can add up to 10–40 homopolymer residues at the end. Commonly, the annealed circles with nicks and mismatching are used directly for transformation as these mistakes are repaired after the recombinant molecule enters into the host.

TRANSFORMATION AND MULTIPLICATION OF CLONES

The central step in a gene cloning procedure is to transfer a recombinant clone generated *in vitro*, into bacteria or any other host. The concept and feasibility of molecular cloning is centred around two principles. Ligation *in vitro* generally yields a population of DNA molecules out of which only some are important. Hence the transformation step should ensure that one cell receives a single plasmid or molecule. This results in separation of each recombination from all the others. Each recipient cell needs to separate from all the others in the population to permit isolation of a clone of cells. Isolation of a clone of cells depends upon the property of the host–vector combination being used. Before transformation we have to keep the host–vector combination and screening methods in mind and select the host, as the host properties are very important in transformation and multiplication. In most of the cases, microorganisms which are proved unlikely to survive in nature are used so that accidental release of genetically modified strains can be prevented.

Today, hundreds of such microorganisms are available. Hence when selecting a host for transformation, three important points must be kept in mind. Firstly the host must supply the factor required for the vector replication and should not contain any elements that inhibit vector multiplication or prevent some screening methods being used. Secondly, the cell should not contain any active restriction enzymes being synthesized, as this will cleave the recombinant molecule. Even the host cells which provide *E. coli* dam and dam methyl transferase are avoided, as they will generate replicated recombinant molecules which cannot be cleaved by some restriction enzymes due to methylation of the recognition site. Lastly, the host should not have any phenotypic character which is similar to the vector. For example, if the vector codes for ampicillin resistance, the host must be susceptible to ampicillin antibiotic in the absence of vector. As a general rule, the host cell should be sensitive to particular antibiotics or toxins and should not harbour any extraneous plasmid (e.g. F^+ *E. coli* cells).

After the host is decided, the second step is deciding the approach to introduce the recombinant DNA molecule. Normally three different approaches have gained wide popularity and acceptance. The first one is uptake of plasmid by chemically treated cells. Most species of bacteria including *E. coli* take up only limited amounts of DNA under normal conditions and have to undergo a chemical pre-treatment before they can be transformed efficiently. Cells which have undergone such a treatment are said to be competent cells. Mandel and Higa developed this process in 1970. In this method, mid-log phase *E. coli* cells are taken and washed with 0.05 M $CaCl_2$ twice, to remove the traces of growth medium. Then Ca^{++} ions bind to the membrane with positive charge. Now as the membrane is surrounded by Ca^{++} ions, DNA molecules bind to them due to the nature of DNA molecule. Now Ca^{++} ions act as a bridge between the membrane and DNA molecule. When cells are subjected to the heat shock in the presence of recombinant DNA molecules, i.e., when the cells are exposed to 42°C for 45–55 seconds, the membrane of the bacteria slightly expands, the negative charge gets weakened and holes are created in the membrane, thus creating a pathway for the migration of the DNA and Ca^{++} ions into the cell.

Once inside the cell, the Ca^{++} ions are expelled out. Of course the biological basis of the techniques is not yet understood. This is the explanation widely accepted. This method is very reliable and works well with most strains of *E. coli*. Recombinant cells prepared by this method yield more than satisfactory results for most standard cloning procedures. The other advantage of this method is the ability to retain high transformation efficiency even when cells stored for months at −80°C are used. The only drawback of this method is lower transformation efficiency, which is not sufficient when the aim is construction of the genomic library. This drawback has lead to the development of a sophisticated method where an electrical shock is given to the cells instead of heat shock. This method is called as electroporation.

In electroporation, the cells are exposed to an electric field of 12 kV/cm² for about a period of 5–10 milliseconds. This shock induces holes in the bacterial cell membrane through which the DNA enters before the holes are repaired. Transformation efficiency frequency will be 100–1000 times greater than that of the chemical method. The only drawback in this method is the requirement of washing the cells repeatedly with water prior to electroporation because traces of metals in the bacterial suspension can cause total explosion of the instrument or a heavy reduction in transformation efficiency.

MULTIPLICATION OF VECTOR IN HOST

Whatever the method being used to introduce the recombinant molecule into *E. coli*, the number of cells that actually take up DNA will be relatively low. In the best conditions only 10^7 transformants are expected. For the maximum propagation of the recombinant clones, the medium chosen for plating the cells should not allow the growth of non-transformed cells at all and secondly it should be able to help us in distinguishing the non-recombinants from the transformants.

The strategy that is used depends on the genetic markers present in the vector. Most of the vectors that are available have similar phenotypic markers such as antibiotic resistance, *lacZ* gene, etc. Several plasmid-cloning vectors carrying genes for antibiotic resistance are plated on the antibiotic media for transformation and recombinant selection. After the transformation, the cells are incubated in growth medium/broth for about an hour before the antibiotic selection pressure is applied, as bacteria require some time to recover from the damage occurred during transformation and also to synthesize the protein/enzymes against the antibiotic before the cells actually encounter the antibiotic on plate. If the cells are plated immediately after transformation, the yield of colonies will be low or we will get zero result. After the incubation period, the cells are plated on the medium for the selection. Normally the plates are plated such that every clone gets enough place to grow and form a characteristic colony. For example, if the volume of cells after the incubation is 1 ml then around 10 plates are plated by taking 100 µl or 0.1 ml of the solution on each plate. If the aim is to isolate a single clone, then the total volume can be plated on one or two plates. If the aim is to generate a genomic library, then the volume is plated in 25 plates or in large plates.

The plates are observed after 18–24 hours of incubation (for bacteria) for identification of the clones carrying the vector with insert. If the clones are identified after this time, there are chances of selecting a wrong clone or bacteria without a vector. As bacteria grow in size, they secrete more amount of enzyme which inhibit the antibiotic. After sometime tiny translucent colonies appear around the colony due to clearing of the antibiotic which are basically non-transformants. Because of the absence of antibiotic, these colonies grow. If the plates are observed after 24 hours, these colonies dominate over the original transformed colonies due to the presence of less DNA as they contain only gDNA, which has to multiply whereas the transformants have to divide the gDNA as well as the vector.

The other problem which is encountered during the multiplication of recombinant clones, is the diffusion of enzyme from non-recombinant clones to recombinant ones. This problem is normally encountered with vectors containing *lacZ* gene or blue–white selection as phenotypic marker. When the plates are incubated for more than 36 hours, the β-galactosidase produced in the non-recombinant bacteria will diffuse to the area where the bacterial clone which is a recombinant one breaks the X-*gal* and stains the bacteria which is a recombinant one, thus giving false positive results.

Hence, the multiplication of transformants is an important step in gene cloning. Any mistake (during any of the step) in this process demands a heavy price either in terms of time or cost.

SELECTION OF RECOMBINANT CLONES

After the multiplication of bacteria with inserts, we get a huge number of clones, which contain inserts of use or interest. In addition, a variety of unpredictable and aberrant recombination can occur, including vectors that have incorporated two or more inserts, and recombination in which the inserts have been altered by recombination within the host cell. Thus transformation and multiplication of clone does not mark the end of gene cloning. It is just one of the steps in recombinant technology.

The task of isolating a desired recombinant from a population of bacteria depends very much upon the cloning strategy that has been adopted and type of vector that has been used. Various methods such as genetic and immunochemical, methods, south-western blotting, nucleic acid hybridization and recombination methods are available. But nucleic acid hybridization with labelled probes is most generally applicable.

Genetic Methods

Genetic methods involve using special characteristics or phenotypes coded by the gene present in the vector. This method becomes very efficient when combined with microbiological techniques. Plasmids and cosmids carry drug resistance or nutritional markers which help in the identification of the recombinant. This method is a prerequisite to differentiate the recombinant vector from non-recombinant vector. Most researchers use blue–white selection or insertional inactivation of a drug-resistance marker as the standard method of selection. If an inserted foreign gene in the desired recombinant is expressed, then genetic selection may provide the simplest method for isolating clones containing the genes, For example, if a gene codes for enzymes which are important for metabolizing some toxin, then clones which carry the genes can be selected by transforming into bacteria which are susceptible to the toxin and then plating them on toxin-containing plates. This method eliminates all other clones which do not have inserts, inserts which code for some other proteins and also the genes which are wrongly inserted.

Immunochemical Methods

Immunochemical detection of clones can be applied, provided the bacteria having the vector code for the protein specific of the insert. This method uses the 3 points of ability of IgG molecules for screening.

1. IgG can bind to different determinants or epitopes on the antigen.

2. It binds to plastic or PVC very strongly and cannot be removed by repeated washing.

3. IgG molecules can be radiolabelled very easily and by *in vitro* method.

The only drawback in this method is that it can be applied only when IgG molecules against the insert coded protein are available. In this method, transformed cells growing on the petri plate are replica-plated onto a fresh plate, as the subsequent steps will kill the cell. After replica plating, the cells are killed or lysed by using chloroform or UV. This step releases the proteins outside. Then a sheet coated with unlabelled antibodies (specific against gene/protein of interest) is placed over the medium such that an antigen–antibody complex is formed and sticks to the sheet. The sheet is removed and incubated in labelled antibodies (specific against the gene of interest). The labelled IgG molecules will bind to the antigen–antibody complex to form a sandwich, in which the antigen is in between the antibody. Positively reacting colonies are detected by washing the sheet and making an autoradiography image. The required clone can be collected from the replica plate. This method is useful in identifying the fusion proteins, where two proteins are fused. Just by changing the antibody specificity, only fusion proteins can be identified.

South-western Blotting

This method is widely used to find DNA-binding proteins coding clones. In this method, the clones are lysed after replica plating. Then the released proteins are absorbed onto the nitrocellulose membrane and incubated with the DNA molecule, which is labelled. Normally the radiolabelled duplex DNA or oligonucleotide contains the sequence for which the DNA-binding protein is specific. Positive clones are selected by autoradiography method. As this technique uses a radiolabelled DNA to detect polypeptide on the nitrocellulose membrane, this method is called as south-western blotting. South-western blotting is widely used to identify the transcript factors from the cDNA libraries.

Nucleic Acid Hybridization Method

This method was developed by Grunstein and Hogness in 1975 and is used to rapidly determine the colony which contains the sequence from hundreds of clones. In this method, bacterial colonies are transferred onto the nitrocellulose membrane from the agar plate and lysed by using alkaline solution. When the bacterial cell wall breaks, the proteins and DNA will stick to the membrane, due to the negative charge of membrane. Then the membrane is placed in proteinase K solution to remove the protein bound to the membrane and DNA. Then the DNA is fixed to the membrane by exposing it to UV rays. After baking, the membrane is exposed to labelled RNA for hybridization. Then RNA–DNA hybrids are monitored by autoradiography. A colony which gives a positive autoradiography result can then be picked from the master plate. A great advantage of the hybridization method is generality. It does not require expression of the insert sequences and can be applied to any sequence, if suitable probe is available.

REVIEW QUESTIONS

1. Explain the methods of isolation of gene of interest.

2. Explain the process of construction of chimeric/rDNA.

3. Explain the process behind transformation under *in vitro* conditions.

4. Explain the process by which you differentiate a clone with insert and without insert.

5. Write a short notes on the following.

    i.   Fragmentation

    ii.   Shot gun method

    iii. cDNA synthesis

    iv. Homopolymer tailing

    v.   Linkers and adapters

20

DNA LIBRARIES

The isolation of genes that encode proteins is often the goal of genetic engineering experiments. In prokaryotic organisms, each of these structural genes have a continuous coding domain in the genomic DNA, whereas in eukaryotes coding regions (exons) of structural genes are separated by non-coding regions (introns). Consequently, different cloning strategies have to be used for prokaryotic and eukaryotic genes. In animals or plants, the desired sequence (target DNA) is frequently (presently in a very small portion), roughly less than 0.02% of the total chromosomal DNA. The problem lies in how to clone and select the targeted DNA sequence. To do this, the complete DNA of an organism is cut with a restriction endonuclease and each fragment is inserted in a vector. Then, the specific clone that carries the target DNA sequences must be identified, isolated and characterized.

CONSTRUCTION OF GENOMIC LIBRARY

The idea of amplifying very large mixtures of DNA fragments and then trying to find the single segment of interest was dramatic.

Now it is almost common to use this strategy. The process of subdividing genomic DNA into clonable elements and inserting them into host cells is called creating a library. A complete library, by definition, contains all of the genomic DNA of the source organism and is called as genomic library. A genomic library is a set of cloned fragments of genomic DNA. The process of creating a genomic library includes four steps (Figure 20.1). In the first step the high molecular weight genomic DNA is separated and subjected to restriction enzyme digestion by using two compatible restriction enzymes. (i.e., *Eco* RI and *Sal* E1). In the second step fragments are then fractionated or separated by using agarose gel electrophoresis to obtain fragments of required size. These fragments are then subjected to alkaline phosphatase treatment to remove the phosphate. In the third step, the dephosphorylated insert is ligated into vector which could be a plasmid, phage or cosmid, depending upon the interest of the researcher. In the last step, the recombinant vector is introduced into the host by electroporation and amplified in host.

In principle, all the DNA from the source organism is inserted into the host, but this is not fully possible as some DNA sequences escape the cloning procedure. Genomic library is a source of genes and DNA sequences. A genomic library is a set of cloned fragments of genomic DNA. Prior information about the genome is not required for library construction for most organisms. In principle, the gDNA, after the isolation, is subjected to RE enzyme for digestion to generate inserts. Depending upon the vector type and genome size of the source, the size of insert is selected. For example, if genome is 2.8×10^6 kb and if it is broken down into fragments of size 4 kb then we require 7×10^5 recombinant clones for the work. If the same genome is broken down into fragments of size 20 kb then we require 1.4×10^5 recombinants. Thus the fragment size will influence the number of clones to be formed, so that there is 99% probability of success when a desired clone is searched.

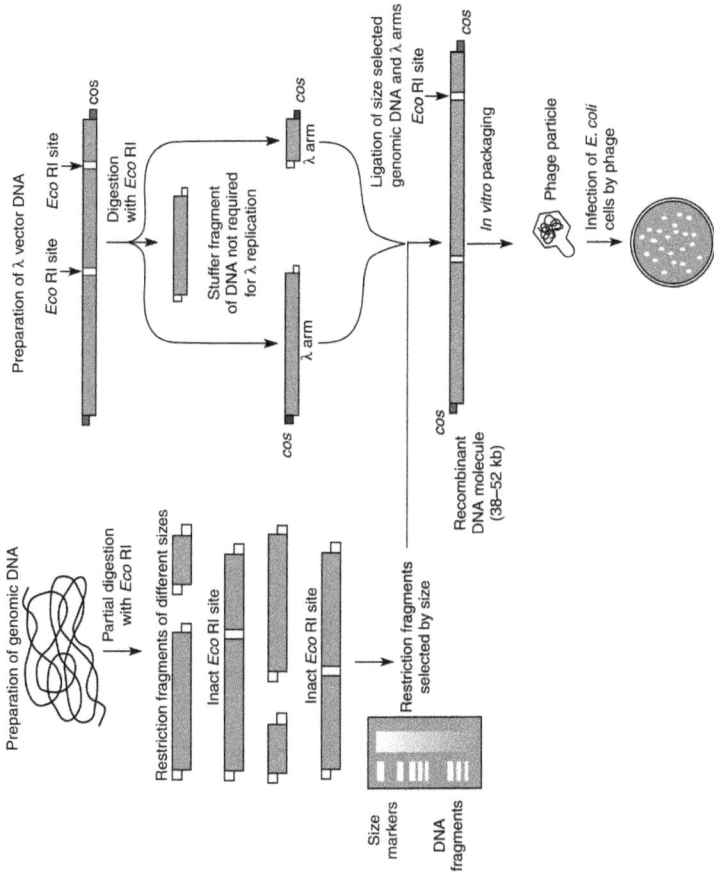

Figure 20.1 Steps involved in construction of genomic library

In reality, the number of clones (f) required for 99% probability is not calculated just by dividing the genome size by size of insert. The number of independent recombinants required in the library must be greater than n. When a library is constructed, some sequences may be over-represented or totally lost due to unknown reasons. To construct an ideal library where all the sequences are equally represented with equal probability, Clarke and Carbon equation is applied. This formula relates the probability (P) of any DNA sequence in a random library of N independent recombinants.

$$N = \frac{\ln(I - P)}{\ln(l - l/f)}$$

where,

N = number of clones required to represent an ideal library

P = probability

f = genome size/size of insert

Large insert vectors have the advantage that libraries containing fewer clones are needed in order to cover an entire genome. Apart from the advantages, they have some disadvantages also. Larger inserts are prone to deletions or rearrangements. Also, the vectors with large inserts are normally maintained at low copy number even when the vector by itself has high copy number. The other problem has the requirement of high-quality, high molecular weight insert DNA, of constant size. Hence, the fragments of size 30–50 kb are generated by R/E, which are then purified by sucrose gradient or by gel electrophoresis. Later these fragments are cloned into vector for amplification. Before ligation to the vector, the digested genomic DNA is treated with alkaline phosphatase to prevent the co-ligation of two inserts (totally unrelated or non-continuous in genome) to a single vector molecule. Even size-fractionation eliminates these artefactual products because they exceed the capacity of the vector.

When a genomic library is constructed, the choice of vector should be carefully considered. Normally phages are used for constructing the genomic library when compared to cosmids and plasmids. Phages can be used for repeated amplification, thus once a genomic library is constructed, it can be used for generations or years together. Secondly, phage recombinants give better and cleaner results when desired clones are identified by colony hybridization. Phages usually give less of a background hybridization than do bacterial colonies. Bacterial populations cannot be stored as readily as phage populations. There is often an unacceptable loss of viability when the bacteria are stored. An important problem in genomic library is "distortion". Not all recombinants in a population will propagate equally well, as the insert size or sequence may affect the replication of a recombinant phage. Hence when a library is put through an amplification step, particular recombinants may be increased in frequency or decreasing or totally lost. With the technological development in genomic library construction methods, most of the researchers prefer to create a new library for each screening rather than take the risk of using previously amplified ones.

Applications of Genomic Library

1. Genomic library construction is the first step in any DNA sequencing projects.

2. Genomic library helps in identification of the novel pharmaceutically important genes.

3. Genomic library helps in identification of new genes which were silent in the host.

4. It helps us in understanding the complexity of genomes.

cDNA LIBRARY

Genomic library is the total representative of total genes present in an organism, whereas a cDNA library is representative of genes, which are expressed during a particular stage of the cell. cDNA library is a collection of clones containing an insert obtained from cDNA cloning. An insert in a cDNA library can express a protein under the presence of a promoter. Genomic library contains introns which are not represented in the mRNA. It also contains inserts, which do not code for any product or protein. Such inserts are obtained from the extragenic DNA. Hence, when the aim of the work is to obtain a gene with protein-making ability or to analyse what genes are expressed to make one cell to convert into another cell type, it is best to go for cDNA library construction than genomic library construction as cDNA library generates a lesser number of recombinant clones, which can be screened very easily.

The first step in the cDNA library construction is the synthesis of cDNA from the mRNA isolated. The mRNA is

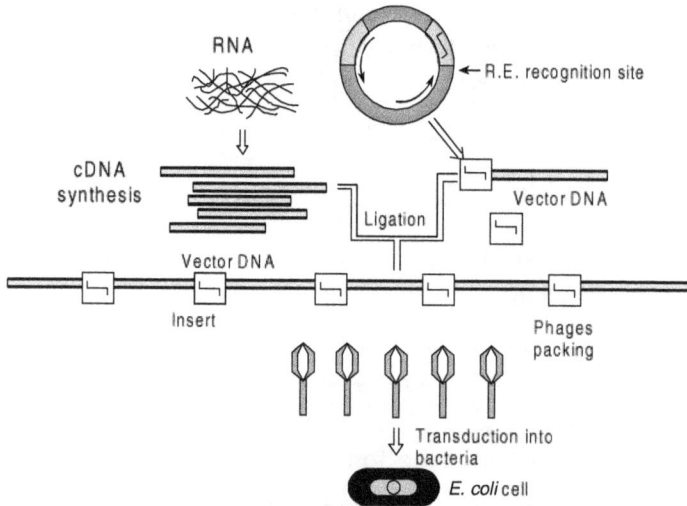

Figure 20.2   Steps involved in the construction of cDNA library

isolated from the total RNA by using biotinylated oligo(dT) method. mRNA is first annealed to oligo(dT), which was biotinylated earlier. This mixture is then mixed with magnetic particles coated with streptavidin. Streptavidin forms a bond with biotin–RNA complex, thus separating it from the other RNA types like tRNA and rRNA. The RNA is then extracted into water and added to a tube containing reverse transcriptase and oligo(dT) (Figure 20.2).

The first step in cDNA synthesis is isolation of the mRNA from the tissue where they are produced in high quantities. These mRNA are placed in tubes containing reverse transcriptase enzyme a small primer or oligo(dT) and dNTPs. The oligo(dT) binds to the mRNA at it's 3′-OH end, as mRNA, especially eukaryotic mRNA contains a poly(A) tail at its 3′-OH end. The oligo(dT) acts as a primer and provides 3′-OH end for the reverse transcriptase enzyme. Reverse transcriptase enzyme adds dNTPs to new complementary DNA by taking RNA molecule as the template strand. In the second step, using RNase, H or alkali treatment, RNA molecule is degraded. Then the newly synthesized cDNA are separated and added to a fresh tube containing DNA polymerase and dNTPs. No primer is required at this step because the variable length of the hairpin loop is formed transiently at the 3′-end of the first strand, which serves as a primer for the second strand DNA synthesis. After the second strand is synthesized, the hairpin structure is removed by using S1 nuclease enzyme.

The cDNA obtained is then ligated to the dephosphorylated adapters so that subsequent cloning becomes easy and efficiency of success increases. In this step the dephosphorylated adapters are used so that the adapters do not ligate among themselves. After the adapter is ligated using kinase enzyme they are phosphorylated. After the phosphorylation, the cDNA molecules are fractionated or separated from the unligated adapters by using column chromatography. Purified cDNA molecules along with adapters are then ligated to the vector molecule and transformed into the bacteria. Phages or bacteria are collected and stored. There are two types of cDNA libraries that are widely constructed.

Selective cDNA Library

Normally during development of multicellular organisms, different genes are expressed (turned on) or suppressed (turned off) in a regulated way. Some genes are expressed only once or in limited cell types or only during some particular stage. As a result, the cytoplasmic mRNA in different tissues and cell types contain distinctive populations of molecules. This differential gene expression can be used to clone cDNAs corresponding to the regulated genes. Even if nothing is known about their gene products, the cDNA library prepared by mRNA isolated from cells present in the gastrula stage and that of muscle cells are different. Each of them represents a different set of genes or clones. Such a cDNA library is called as selective cDNA library.

Expression cDNA Library

Expression cDNA library is constructed when we want to know the proteins that are made during a particular stage of cell. cDNA molecules generated are cloned into vectors that are specially designed to permit transcription and translation of the cDNA coding region. The clones thus produce the protein of our interest or particular to the cell from which the mRNA is isolated. The desired clone can be identified by immunological screening or by non-stringent hybridization method. This method is helpful in cloning when nothing is known about the structure of the gene or the protein.

REVIEW QUESTIONS

1. Write the principle involved behind genomic library construction.

2. Write in detail the principle behind cDNA library construction.

3. List the difference between genomic and cDNA libraries.

4. Explain the difference between selective and expression cDNA library.

21

POLYMERASE CHAIN REACTION

PCR stands for polymerase chain reaction and sometimes it is also called as people choice reaction. PCR was devised by Kary Mullis in the mid 1990s and it revolutionized the field of molecular biology by making possible a whole set of new approaches towards the study and analysis of a gene. PCR is an extremely powerful new procedure that allows one to amplify a selected DNA sequence from a genome a million-fold or more in matter of hours.

The concept of PCR was envisaged by Harghobind Khorana in early 1970, but due to lack of knowledge about DNA polymerase and gene synthesis technique, the idea was dropped. Later after 15 years the idea was independently conceived by Mullis and he brought PCR from the stage of idea to wide applicability. Now PCR has become a fundamental cornerstone of genetics, for basic science, etc. It will not be an exaggeration if we say that there is no field in science which is not influenced by PCR. The main reason for its success is simplicity, flexibility and robust speed of gene application.

Components of PCR

Polymerase chain reaction, like other chemical or biological reactions, requires various components. A basic PCR requires several important components.

Taq polymerase   *Taq* polymerase is the DNA polymerase isolated from a thermophilic bacterium called *Thermus aquaticus*. This bacteria lives in water at a temperature of 75°C and its DNA polymerase is very stable up to 100°C and functions efficiently at 72°C. DNA polymerase isolated from *Thermus aquaticus* is called as *taq* polymerase. The only problem with *taq* polymerase is the absence of the proofreading capability. This enzyme incorporates one incorrect nucleotide for every 2×10^4 base pairs that are added. But this problem is solved using protein engineering. Some commercial companies now supply *taq* polymerase with proofreading ability.

Primers   Primers are the most essential components of PCR. Without primers there is no PCR. Primers are single-stranded DNA molecules or oligonucleotides complementary to the starting and ending sequence of gene or DNA to amplify. Primers are normally 18–30 bp in length. The primer complementary to the starting sequence of gene or DNA to amplify is called as forward primer and the primer complementary to end sequence is called as backward primer. The forward and backward primers are never complementary to one another. Always primers contain 40–60 G/C nucleotide. Primers do not contain a long run of single nucleotide (i.e., 5′ GCAAATTAAAAAAA) and they do not form secondary structure.

Theoretically, primers are short, single-stranded DNA sequences synthesized artificially as per the user requirement and adhering to the basic principle of base pairing.

dNTPs   PCR requires equimolar amount of dATP, dTTP, dCTP and dGTP. dNTPs stand for deoxynucleotide triphosphates.

These dNTPs are used by the *taq* polymerase during amplification and are incorporated into newly synthesized DNA.

Bivalent cations   As *taq* polymerase requires Mg^{++} as cofactor for activity, it is normally used in PCR at the concentration of 1.5 mM routinely. But in some cases, when there is a problem of non-specific attachment of primers, the concentration is kept a bit high, i.e., at 4.5 mM to 6 mM.

Template   Template contains target sequence and can be added to PCR as a single strand or double strand. Template can be a DNA or RNA (for RT-PCR, i.e., Reverse transcription PCR). It can be circular or linear. Linear DNA are amplified efficiently when compared to circular.

Buffer   To maintain pH during the amplification, tris HCl buffer (8.3–8.8) is added. The release of pyrophosphate by the lysis of dNTPs will decrease the pH from neutral to acid. Thus tris HCl prevents this shift and maintains the pH at 7.2. Even the incubation of the reaction mixture at 72°C make the pH of any buffer to fall by one unit, hence tris HCl at pH 8.3 is used.

Monovalent cations   Standard PCR buffer contains 50 mM KCl, as KCl enhances the amplification.

BASIC REACTION

PCR exploits certain features of DNA replication. DNA polymers use single-stranded DNA as template for the synthesis of a complementary new strand. In DNA replication, single-stranded DNA is generated by DNA gyrase enzyme, but in PCR the double-stranded DNA is converted into single-stranded form by using heat, (i.e., by raising the temperature to boiling point). DNA polymerase requires a free 3′-OH for extension. In replication a small DNA is synthesized by primers. But in PCR this need is fulfilled by the primer. A typical PCR procedure involves a number of cycles of three successive steps for amplifying the DNA (Figure 21.1). All the components required for PCR are mixed in a small eppendorf and a layer of mineral oil is placed over it to prevent the loss of water during the PCR by heating. As the loss of water increases, the concentration of components increases, thus leading to mismatching and improper amplification. One round of denaturation, extension and annealing is called as thermal cycle.

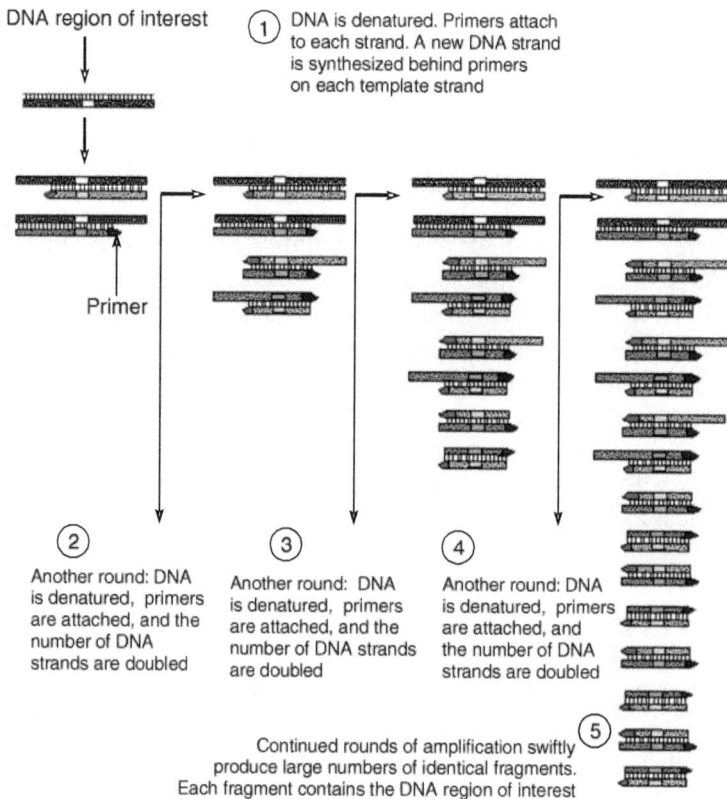

DNA region of interest

1 DNA is denatured. Primers attach to each strand. A new DNA strand is synthesized behind primers on each template strand

Primer

2 Another round: DNA is denatured, primers are attached, and the number of DNA strands are doubled

3 Another round: DNA is denatured, primers are attached, and the number of DNA strands are doubled

4 Another round: DNA is denatured, primers are attached, and the number of DNA strands are doubled

5 Continued rounds of amplification swiftly produce large numbers of identical fragments. Each fragment contains the DNA region of interest

Figure 21.1   Polymerase chain reaction

Denaturation   This is the first step in PCR amplification. In this step the temperature of DNA sample in the eppendorf is raised from room temperature to 94–96°C and kept at that for 5

min., so that the double-stranded DNA is converted to single-stranded. 95°C temperature is maintained for 5 min. only in the first cycle. From the second cycle, the temperature (95°C) is maintained for 20–40 seconds only.

Annealing This is the second step in PCR. During this step, the primer binds to the DNA, complementary to the sequence (and forms the hydrogen bond with template DNA and have 3′-OH protruding out freely). This step is facilitated by lowering the temperature to 55–58°C for 1 min. The temperature maintained in this PCR is a key variable which determines the success and specificity of PCR. The temperature of annealing varies from PCR to PCR. This step is also called as renaturation.

Extension This is the third and last step in PCR. The temperature is raised to 72–76°C as it is the optimum temperature for the *taq* polymerase to act. *taq* polymerase at its optimum temperature use 3′-OH of primer and template strand as complementary strand and synthesize the new strand. This step usually lasts for 3–5 min. depending upon the length of target site to amplify.

The most important and significant aspect of PCR is the generation of amplified DNA fragments of a fixed length. This happens because the primers have the ability to bind to target DNA as well as to the amplified DNA fragments. During the extension step in the first thermal cycle, the primer extends beyond the site of the second primer. These new strands form a "long template" that will be used in the second cycle. The primer hybridizes both to the new template and to the template strand DNA in the second cycle. This cycle produces long templates as well as some strands that have a primer sequence at one end and a sequence complementary to the other primer end thus producing a short template.

During the third cycle, short template, long template and original DNA template will hybridize with primers and are amplified. In subsequent cycles, the short template preferentially accumulate and by the 30th cycle there are about a million strands.

APPLICATIONS OF PCR

PCR finds immense applications which are as follows:

1. *Amplification of specific DNA sequence*   Specific DNA, which is our interest, can be amplified just by using primer against our target sequence. For example human insulin gene can be amplified by using primer which binds to the insulin gene at first end of the gene.

2. *Single-strand DNA for sequencing*   Single-Stranded DNA can also be made by PCR, if we use different concentration of primers. In this process the concentration of forward primer is more than backward primer (i.e., 100-fold). First, amplification takes place by using both the primers then once the backward primers get exhausted, only one primer continues to anneal, thus generating single-stranded DNA.

3. *Sequencing PCR*   Nowadays PCR is used for DNA sequencing. The DNA is amplified along with the component and with dideoxynucleotide so that smaller DNA fragments with one nucleotide difference are generated. These fragments are separated by gel and analysed by chromatogram.

4. *Detecting mutations*   Mutations occur in cancer cells and in inherited disorders. Knowing the nature of the mutation in a patient is important for diagnosis and therapy. PCR amplification is proving invaluable as a tool for screening particular genes, e.g. *ras* gene, for mutation.

5. *Detection of infectious agent*   Conventional methods including growing the organism in culture or detecting antibodies

are difficult and time-consuming. PCR amplification can be performed by using primers, which are specific for genes present only in the infection agent.

6. *Sex determination* Genetic analysis of a smaller number of cells is very important in prenatal diagnosis. Sex determination is the fist step in prenatal diagnosis so that the probable disease list that the foetus can possess can be finalized. Sex determination is carried out by using primers against unique sequences on the Y-chromosomes. If amplification occurs, the foetus is male, if not it is female.

7. *Molecular evolution* Mitochondrial genes do not undergo rearrangement during meiosis, and are transferred via the maternal route and undergo a mutation at a faster rate. Thus amplification and sequencing of mitochondrial genes from preserved fossil will throw light on the process of evolution and divergence. It is very easy to amplify mitochondrial genes by using PCR.

DRAWBACKS OF PCR

An unwelcome corollary of the amplification power of PCR is that minor contamination of the starting material can have serious consequences.

PCR can amplify DNA from living cells and a dead cell thus putting a doctor in confusion whether the person is having a dead or living infectious agent in the case of absence of symptoms.

REVIEW QUESTIONS

1. Write in detail about the PCR components.
2. Explain the use of Tris buffer of pH 8.3 in PCR.
3. Give an account of the applications of PCR.

4. Write short notes on the following.

    i.   *taq* polymerase

    ii.   Primers

    iii. Bivalent cations

    iv. Template

    v.   Buffers

    vi. Drawbacks of PCR

    vii. Application of DNA sequencing

22

DNA SYNTHESIS BY CHEMICAL METHOD

DNA can be synthesized artificially in a test tube by chemical process instead of using biological synthesis. The process of synthesizing a short stretch of DNA by using a chemical reaction is called as chemical DNA method. Machines that automate the chemical reaction involved in DNA synthesis which can synthesize a single strand, when the sequence and chemicals are provided, are called as gene machines or DNA synthesizers. A DNA synthesizer consists of a set of valves and pumps that are programmed to introduce specific nucleotides and the reagents required for the coupling of each consecutive nucleotide to the growing chain. Chemical DNA synthesis does not follow the biological direction of DNA synthesis. In the chemical process each incoming DNA is added to the 5′-hydroxyl terminus of the growing chain. These processes of adding are controlled by computer and are carried in a single reaction vessel that is a column, so that reagents from one reaction step can be readily washed away before the reagents for the next step are added and the reagents can be used in excess to drive the reactions as close as possible to compilation.

PHOSPHOROAMIDITE METHOD

The phosphoroamidite method of chemical DNA synthesis is widely used. Nucleotides that are going to be used in the synthesis are normally modified or blocked at various sites in the molecule so that the side reactions are prevented during chain growth. The amino groups in the adenine are derivatized by using benzoyl, whereas guanine and cytosine are derivatized by using isobutyryl and benzoyl groups respectively. Thymine is not treated because it does not have an amine group. The 3′-OH of the nucleotides are blocked by using diisopropylamine group. Diisopropylomine group has a methyl group at its 3′-phosphite group. 5′-PO_4 of the nucleotide is blocked by using dimethoxytrityl (DMT) group. Every possible site of non-specific cross-reaction is blocked before starting the extension or synthesis. This structure where a nucleotide is blocked by DMT (at 5′-end) and diisopropylamine group with methyl group is called as phosphoroamidite.

The synthesis cycle of DNA starts with the attaching of the first nucleoside to the inert solid support (which is a controlled pore glass bead in most of the cases). The nucleoside is attached to the spacer by using 3′-OH of the nucleoside. After the attachment, the column is washed with acetonitrile to remove water and unbounded nucleotide. Later the column is washed or flushed with argon to remove the acetonitrile. The TCA (trichloroacetic acid) is added into the column to remove DMT from the nucleotide attached to the spacer. The column is once again washed with acetonitrile to remove TCA and with argon to remove acetonitrile. This step is called as detrilylation. The next nucleotide in the form of phosphoroamidite is added along with tetrazole simultaneously. Tetrazole activates the phosphoroamidite so that its 3′-phosphite forms a covalent bond with the 5′OH group of the initial nucleoside. A covalent bond that is formed is in the form of a phosphate triester bond.

The column is flushed with argon to remove unincorporated phosphoramidite and tetrazole. As phosphoroamiditedoes not bind all the support-bound nucleosides, during the first coupling reactions, the unlinked residues must be prevented from linking to the next nucleotide during the following cycles. Hence, the column is washed with acetic anhydride, and dimethyl aminopyridine is added to acetylate the unreacted 5′-OH groups. This step of coupling is required to prevent the synthesis of fragments with different lengths and sequences.

In the next stage, the column is flushed or washed with iodine mixture ($KI + I_2$) for oxidation of the phosphate triester to form pentavalent phosphate triester. This step is required because phosphite triester bond is unstable and is most likely to break in the presence of either acid or base. This step is called as oxidation.

The process or cycle of detritylation, phosphoroamidite activation, coupling, capping and oxidation are repeated until the last nucleotide is added to the growing chain. The newly synthesized DNA are bound to the solid inert support with methyl group on the pentavalent phosphate triester bounded amino group. The last nucleotide that is added contains a DMT at 5′ end, which is removed by TCA. Hence the last nucleotide does not have 5′-PO_4. From the DNA molecule, benzoyl and isobutyryl are removed and DNA nucleotide is detritylated. Before the detritylation, a phosphate group is added at the 5′ end with the help of alkaline phosphatase and ATP. Then the DNA molecule is cleaved from the spacer molecule leaving a 3′-OH terminus. From the column the DNA is eluted and purified by using either reverse-phase high-pressure liquid chromatography or by gel electrophoresis.

Application

There are immense applications of chemically synthesized DNA molecules or oligonucleotide.

1. Single-stranded oligonucleotides are used as probes in medical research, as primers in PCR, in sequencing and in *in vitro* mutagenesis.

2. Double-stranded oligonucleotides can also be synthesized from single-stranded DNA by using DNA polymerase and a short single-stranded primer.

3. These double-stranded oligonucleotides are usually used as adapters and linkers.

4. Genes with effective set of codons that are better suited to the host organism can be synthesized without changing the true amino acid sequence of the protein.

Drawbacks

The major drawback in this method is that the process is time-consuming and involves excessive wastage of phosphoroamidite molecule. Secondly, as the length of the oligonucleotide increases the yield falls down.

DNA SEQUENCING

Nucleotide sequences contribute to the gene or chromosome. A change in the sequence or nucleotide pair changes or alters the function of the gene. The central core of genomic science or genomics lies in the understanding and elucidation of the nucleotide sequences present in the genes.

Until 1970, the idea of gene sequencing was barely conceivable. But today the chromosomes of at least 100 different organisms, have been sequenced entirely. Millions of individual genes are also sequenced. This drastic change from an unconceivable idea to wide practical utility occurred due to the development of three important techniques—gene amplification techniques, which provide large quantities of DNA fragments, improvement of gel electrophoresis procedure, which can separate DNA molecules, which are different in single base pairs.

Till now, three methods of large-scale DNA sequencing have been developed. Of all of them, the automated DNA sequencing method is the latest and is presently used widely all around the world. Maxum–Gilbert method is the method that was developed first and uses chemical reactions. But all the methods of DNA sequencing depend on the generation of DNA fragments with one end having a common or fixed nucleotide. For all the fragments that are generated, the other end has variable or different nucleotides.

Maxum-Gilbert Method

This method was developed by A. Maxum and W. Gilbert in 1977. This method uses different chemicals to break or alter the DNA at different points or bases in the DNA.

The process involves radiolabelling of the $5'$-PO_4 of the DNA molecule to be sequenced by using alkaline phosphatase and ATP or using a chemical method. The DNA molecules are divided in four groups or transferred into four tubes. To the first tube, dimethyl sulphate is added, which methylates guanine at the N 7 position. To the second tube, acid is added, which alters either adenine or guanine. Hydrazine is added to the third tube, which alters either thymine or cytosine. To the fourth tube, hydrazine along with NaCl is added to alter cytosine base pair. A fifth tube is kept as a reference containing only NaOH. The fifth tube gives a redundant, but confirmatory inform.

Piperidine is added to all the tubes to remove the altered base pairs and break the DNA at the sugar residue where the base is removed. Chemicals are added such that base-specific reactions are carried out at one site per molecule. This sequencing reaction is analysed by running the four or five samples side by side on a sequencing gel. A sequencing gel is a high-resolution gel designed to fractionate single-stranded (denatured) DNA fragments based on their length. These gels contain 7M area and 6–20% polyacrylamide. Electrophoresis is carried out at 70°C to prevent the formation of secondary structure. After electrophoresis, the gel is exposed to large sheets of X-ray film in dark for 24 hours. Radiolabelled phosphate present on the DNA strands or fragments emit energy, which reacts with the X-ray film to produce a band corresponding to the location or position on the gel. This technique of generating bands by exposing X-ray film to radioactive material is called as autoradiography. The film is developed and the sequence is read directly from the sequencing ladder in adjacent base-specific tracks.

Chain Termination Dideoxy Procedure

This method is now used for large-scale DNA sequencing. This procedure is developed after understanding two factors—the ability of DNA polymerase to synthesize a single-stranded DNA template (M13 phage) and the ability of T7 DNA polymerase to incorporate dideoxynucleotide triphosphate in the chain randomly during the DNA synthesis. Dideoxynucleotide (ddNTP) is analogous or similar to the deoxynucleotide triphosphate except that it does not have 3′-OH at its 3′-carbon position in the sugar (Figure 22.1). In practice, four reaction mixtures are set up, each containing all the four normal deoxynucleotides (dNTP) and one dideoxynucleotide. In a given reaction, the ratio of dNTP and ddNTP is kept at approximately 100:1, so that there is only one chance of termination at a given point in the nascent chain.

All reaction mixtures contain a template strand, a single primer (usually forward primer) *taq* polymerase and components of PCR. PCR reaction is carried out, but the only difference here is that the fragments that are generated are not of uniform length, but of

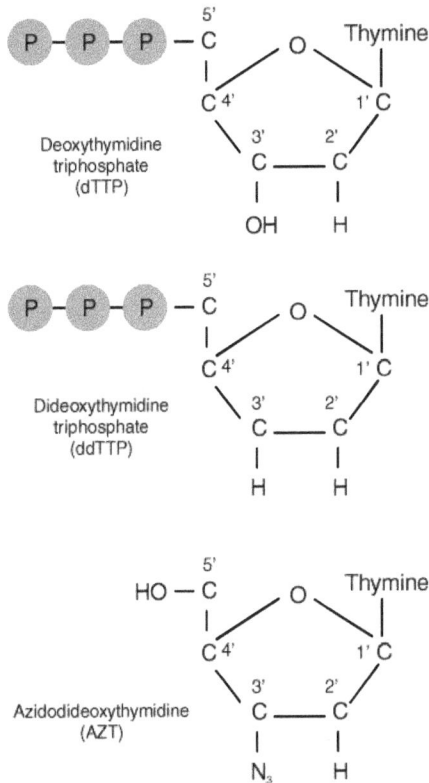

Figure 22.1 Molecular structure of ddTTPs used in DNA sequencing

varying length. This occurs because wherever a dideoxynucleotide is added, the extension step stops because the next nucleotide added cannot be added to the dideoxynucleotide, as the phosphodiester bond is not formed. After the PCR reaction is completed, the template is separated from the newly synthesized strand by denaturation and separated by polyacrylamide gel electrophoresis. The position of each bond is detected by autoradiography. The shortest fragment will migrate the greatest distance (i.e., nearer to the anode (or positive) electrode). Each successive band will contain chains that are one nucleotide longer than the chain in the preceding band of the ladder. The DNA sequence can be deduced by reading the ladder produced by autoradiography of the polyacrylamide gel used to separate the fragments. Currently by this method, DNA of length 100 bp can be read, but the sensitivity and length of DNA fragment that can be sequenced can be increased by using ^{35}S-'deoxynucleotide triphosphate. ^{35}S-dNTP contains a sulphur atom at one of the oxygen position phosphate nucleotide molecule. The chain termination DNA sequencing method is shown in Figure 22.2.

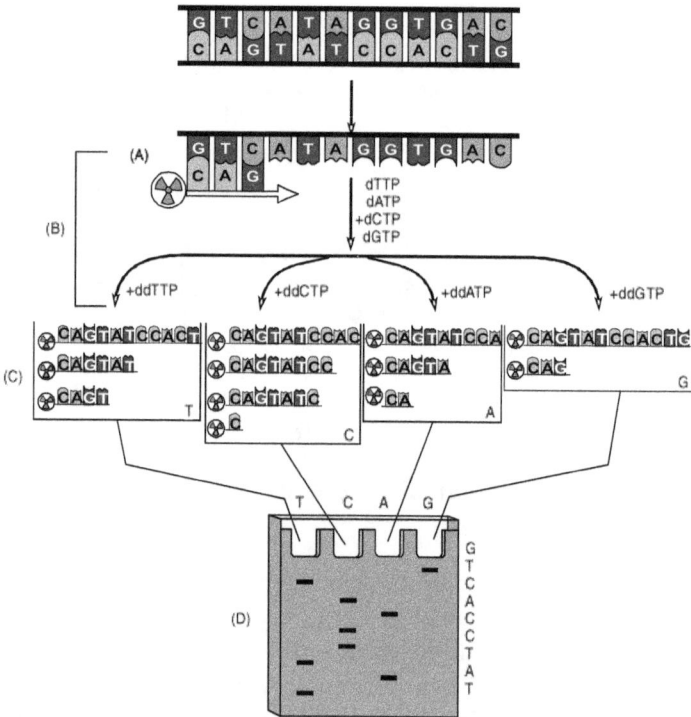

Figure 22.2   Chain termination DNA sequencing method

Automated Sequences

In manual sequences, the DNA fragments are radiolabelled in four different reaction vessels separated on the gel in four different lanes and detected by autoradiography. This is cumbersome and time-consuming making them an invaluable tool to be used in large-scale sequencing projects. To overcome this problem, automated DNA sequencing method is developed (Figure 22.3). This method is a more modified version of dideoxy method. In this process, the dideoxynucleotide that is going to be incorporated in the DNA molecule is not radiolabelled but has a chromogen molecule. Hence, the fragments that are generated have a colour or a chromogen molecule at the end of the chain.

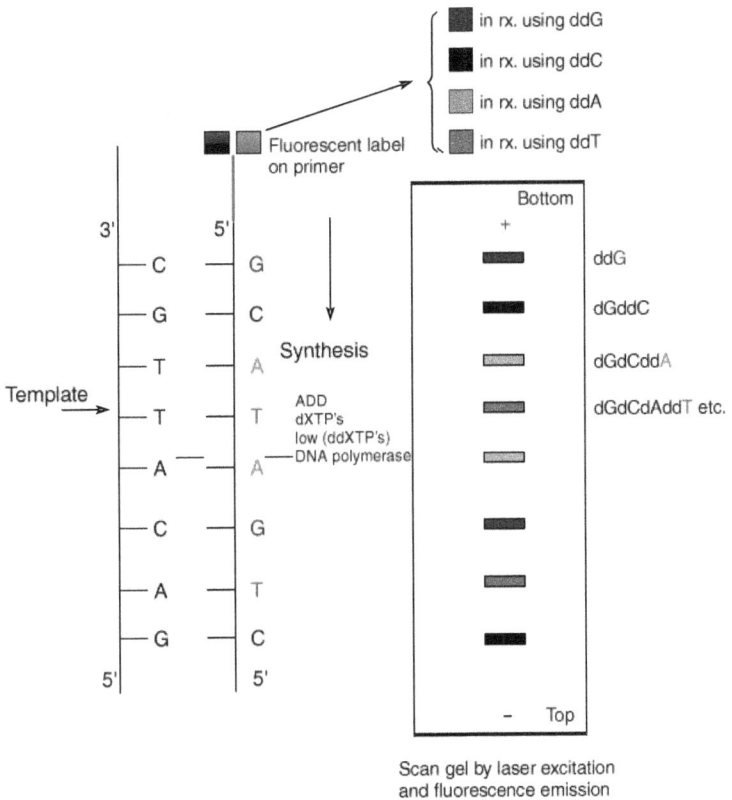

Figure 22.3 Automated DNA sequence

Each dideoxynucleotide is attached by a different fluoromolecule so that the nucleotide at the chain termination can be identified, (i.e., all dGTPs have a green colour emitting fluoromolecule attached to it and all dQTPs have a red colour emitting fluoromolecule, etc.). The other difference is that all the reactions are carried out in the same reaction tube and are run in a single lane in polyacrylamide electrophoresis rather than in four lanes.

As there is no radioactive material in the reaction there is no question of using autoradiography for sequencing. To deduce the sequences, the DNA bands are detected by their fluorescence as they electrophorese past a detector. The molecule or nucleotide at the end can be identified by identifying the flurogen, as we know what flurogen is attached to which nucleotide. The sequence that is generated in this method will be of complementary strand, as the DNA sequencing proceeds from $5' \rightarrow 3'$ end and follows Watson–Crick base pairing. Hence the sequence that is obtained is the complementary strand. In automated sequencing, the entire process of detecting and deducing the nascent template DNA sequence is carried out by computers.

Application of DNA Sequencing

1. DNA sequencing information is important for planning the procedure and method of gene manipulation.

2. DNA sequencing is used for construction of restriction endonuclease map.

3. It is used to find tandem repeats or inverted repeat for the possibility of hairpin formations.

4. The sequences can be used to find whether any open reading frame (ORF) coding for a polypeptide exists.

5. DNA sequences can be used to find a polypeptide sequence from the data bank or to compare with DNA sequences from other animals for phylogenetic analysis.

6. They are used to construct the molecular evolution map.

7. They are useful in identifying exons and introns.

REVIEW QUESTIONS

1. What is DNA synthesizer?

2. What is the fundamental difference between biological and chemical method of DNA synthesis?

3. What is phosphoroamidite molecule?

4. Explain the steps involved in chemical DNA synthesis.

5. Write about Maxum–Gilbert method of sequence.

6. What is the difference between dideoxy and automated DNA sequencing method?

23

MOLECULAR MARKERS

RESTRICTION FRAGMENT LENGTH POLYMORPHISM

The presence of a given molecule in multiple forms is termed as polymorphism. Restriction fragment length polymorphism, widely called as RFLP, simply means the generation of fragments of different lengths when it is cut with a restriction enzyme. This method was invented by Alec Jeffreys.

The procedure of RFLP involves 3 basic steps (Figure 23.1). First, the total genomic DNA is isolated from the cell and subject to a restriction enzyme digestion. These restriction digests will be electrophoresed to separate the DNA fragments based upon the size. These DNA fragments are then transferred on to a nitrocellulose or nylon membrane by using Southern blotting. Then by using a specific probe we can try to hybridize it with Southern blot to generate a band pattern. The principle behind the RFLP technique is based upon the ability to distinguish polymorphic alleles of a given DNA sequence. In humans, mutations occur at a high rate especially the single base-pair mutations. These mutations are insignificant if they do not alter the protein structure. However, some mutations may altogether abolish a restriction enzyme recognition site or create a new enzyme. Hence, we get a new gene which does not have the RE recognition site. We all know that there will be two or more copies of genes present on the two chromosomes. These genes are similar in sequence and function with small variations. Such genes are called as polymorphic genes.

The second significant contributors for polymorphism of a gene are the unequal recombinations that occur during meiosis. Normally at the end of the chromosomes, there are repeats of DNA sequences that are adjacent to each other without any intervening DNA sequences. Such sequences are called as tandem repeats or microsatellites. At a given loci, one chromosome may contain 2–20 tandem repeats and at the same loci on the other homologous chromosomes, there may be 5–15 tandem repeats. Such repeats are called as variable number of tandem repeats (VNTR). For example let us suppose that the VNTR is 500 bases with three restriction enzyme recognition sites and generates fragments of length 100,150 and 250 bp in length. If one of the base pairs is mutated in the recognition site-II (which generates 150 bp length fragment) and the restriction enzymes fails to recognize this site, we will get only two fragments of length 100 and 400 bp. Assume that this mutation has occurred in only one gene present on one chromosome and is absent on the other chromosome. Then when both the genes are subjected to the restriction enzyme digestion, we will get four different fragments of lengths 100, 150, 250 and 400 bp. This pattern of band generation is specific for every individual and is called as DNA fingerprint. When a probe is used to hybridize, it will bind at four sites generating four bands.

As a gene exists at various loci on the chromosome, sometimes some tandem repeats are inserted into the gene at random sites thus increasing the length of the gene, which in turn alters the length of the fragment. The gene will be located at different points on the gel. When the probe hybridizes, it will bind at all these different points and generate a band. As there is only a very small chance of alterations or mutations occurring in the same gene for two persons, the band formed will be different. Normally a single probe will be able to find 50 different variable sites.

Thus in a population of individuals, two possibilities exist for a single chromosome with respect to this site—it can be cleaved (+) or not cleaved (–). The actual analysis of DNA samples from a group of individuals is slightly more complicated

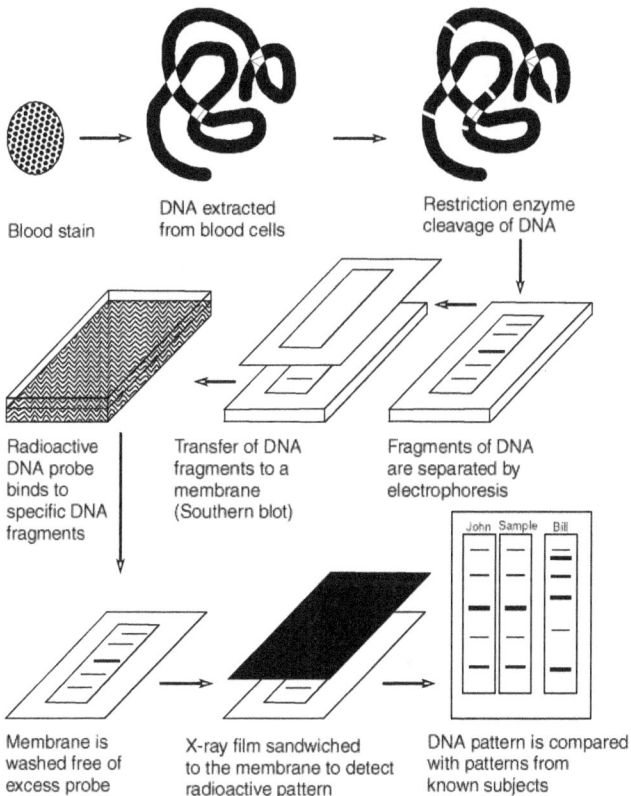

| Blood stain | DNA extracted from blood cells | Restriction enzyme cleavage of DNA |

| Radioactive DNA probe binds to specific DNA fragments | Transfer of DNA fragments to a membrane (Southern blot) | Fragments of DNA are separated by electrophoresis |

| Membrane is washed free of excess probe | X-ray film sandwiched to the membrane to detect radioactive pattern | DNA pattern is compared with patterns from known subjects |

Figure 23.1 RFLP procedure

because a chromosome occurs as pair. However each genotype (+/+, +/– and –/–) produces a distinctive pattern of fragments after hybridization with the probe (Figure 23.2).

When a single site is examined, there are two possible haplotypes (+/–). The genetic status of each RFLP locus on a single chromosome is called the haplotype. Thus when two different sites on a chromosome are examined, there will be 2^2 or four haplotypes. When n number of loci on a chromosome are studied, 2^n haplotypes will be available.

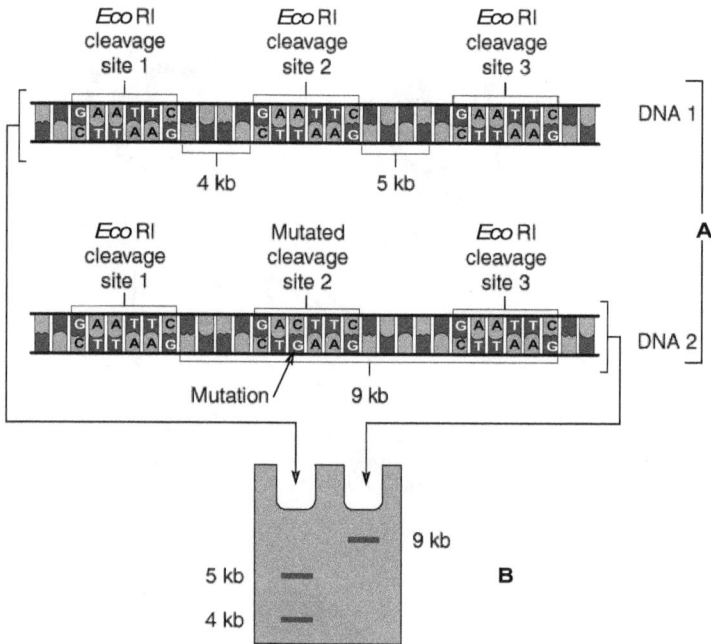

Figure 23.2   Molecular basis explanation for difference in RFLP pattern for two different individuals

The determination of which alleles of RFLP loci are present in the chromosome of an individual is called as haplotyping, genotyping or DNA typing. RFLP loci are inherited in a strictly

Mendelian manner of inheritance, and the inheritance of a locus can be traced with the pedigree. In practice, DNA from a number of individuals is treated separately with different restriction endonucleases and then tested with cloned single copy DNA segments as probes to detect RFLPs.

Much of the human genome cannot vary greatly between individuals because it has an essential coding function. In non-coding regions, this requirement does not exist and thus the DNA sequence can accommodate changes. Hypervariables are those that differ greatly between unrelated individuals. This polymorphism is due to the variation in the number of repeats arising from loss or gain of the repeated sequences through mutation.

Each hypervariable locus has some sequence similarity with many other loci. Microsatellite is a class of hypervariable loci consisting of tandemly repeated, short, simple sequences of 100 to 200 bp length distributed in eukaryotic genomes. Microsatellite DNA is absent in prokaryotes. DNA fingerprints are very complex. They contain dozens of bands, some of which smear together, making them hard to interpret. Hence probes that hybridize to single polymorphic DNA are developed. A set of four such probes are used to give a definite band pattern.

Applications

1. RFLP technique is widely used to identify the father in the case of paternity disputes. As we know, every individual in the world will receive half set from the biological mother. If at all the child is the progeny of the parents, then it will inherit the RFLP loci pattern (at least 70%) without alterations. Hence when bands are generated, they should match with father and mother bands of RFLP or DNA typing.

2. RFLP technique is used in forensic science to confirm whether the acquitted has committed the crime especially when there is no eye witness. All human beings shed their skin cells. Hence, when a crime is done, there is always a possibility of the culprit leaving some cells such as hair, blood or semen in the case of a rape. These samples provide enough DNA for RFLP analysis, as the DNA is almost fixed. If the band pattern of the suspect and DNA fingerprint obtained from the DNA collected from crime site are similar, one can be convinced that the suspect has committed the crime.

3. The other area where RFLP is widely used is in the identification of culprits involved in wild life poaching (killing animals which are protected by wild life acts). Most poachers chop the animal into pieces and try to smuggle it in the name of chicken/lamb meat, etc. To identify the meat when doubtful, RFLP is carried out using probes which specifically bind to only one type of animal cell. If a probe (e.g. specific for deer) binds to the RFLP generated from the meat collected from suspect, it means that the meat he is carrying is deer's meat.

4. RFLP technique is not only used in forensic science, but for diagnostic purposes also. It is widely used to identify sickle-cell anaemia in neonatals. In sickle cell anaemia, a single base pair mutation occurs at the 18 bp, leading to coding of glutamic acid instead of valine. This alteration not only creates a shift in amino acid, but also eliminates the restriction enzyme recognition site of Mst II. Hence when the DNA from a normal homozygous person is digested with Mst II and probed with β-globin cDNA two bands will be generated. But in the case of a sickle-cell anaemic patient, only one band is generated as the β-globin gene is not cleaved into two fragments of length 0.2 kb and 1.1 kb. But in the case of a heterozygous person (with one normal gene and other for sickle cell anaemia), three bands

are generated corresponding to the size 0.2 kb, 1.1 kb and 1.3 kb.

RANDOM AMPLIFIED POLYMORPHIC DNA (RAPD)

Random amplified polymorphic DNA (RAPD) is widely used to analyse the plant DNA. This is because plants use the self pollination mechanism and so the RFLP patterns cannot be applied as effectively as for animals. For any crop improvement programme, as the first and foremost step is to have a reliable estimate of genetic relationship and genetic diversity, a large number of polymorphic markers are required. RADP method can provide enough markers in an easy, cost-effective manner and in minimum time. In this procedure (Figure 23.3), PCR reaction will be carried out on the DNA sample isolated from the plant. In this process, primers that are used are 9 to 10 bp in length containing 50–80% of G/C content and do not have any palindromic sequence. The primers are similar in sequence except in orientation (5´ GGAACGGCAG3´ and the other primers 3´GGAAGGCAG5´), hence a single primer binds to the target DNA.

Whenever a primer hybridizes to both the strands of DNA in the proper orientation, and the two sites are about 100 to 3000 bp apart, the intervening DNA region will be amplified. As we have already studied, DNA fragments generated during amplification have a fixed length and so when the PCR fragments are electrophoresed, they produce bands of fixed size or are seen as bands. The number of DNA fragments in the sample that are amplified depends on the primer and the genomic DNA used. Each time the same primer is used with the same target DNA, the amplified procedure will be the same. If a single nucleotide substitution occurs in the primer sequence, then RAPD patterns that are generated will change.

Hence by using a single sequence of primer or universal primer, a set of plants are analysed. By using a single set of primers consisting of 5–12 different primer sequences, PCR amplification is carried out, thus generating different band patterns.

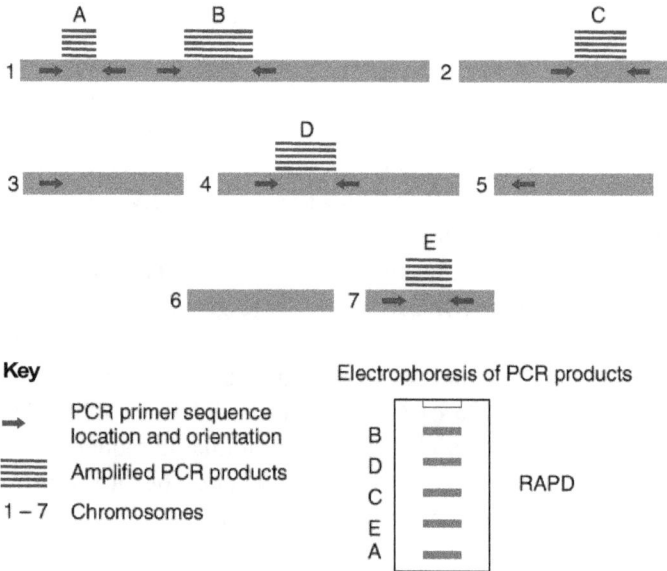

Figure 23.3   Procedure for RAPD

RAPD is used widely because

1. A single set of primers can be used for all plant species.

2. No hybridization or radioactivity is required.

3. The process is simple.

For example, let us suppose that three different primers (from a single set of primer group) have to bind to DNA of all the plants and amplified. They generate a different band pattern. The first primer binds to all the plant DNA and amplifies the DNA of identical length. The second primer binds only to DNA from plant 3 and 1 but amplifies the DNA of varying length. Now, by using three primers we can identify these three plants however similar they are in appearance, as different plants.

RAPD has wide applications in

1. Analysing the genomes

2. Identification of genetic differences among the plants

3. Providing a means of identification of plant variety raised by a cultivator and thus helps in development of specific molecule identification.

4. Providing information to solve problems related to plant breeders' rights.

REVIEW QUESTIONS

1. Explain the theory behind RFLP technique.

2. List the applications of RFLP technique.

3. What are the applications of RAPD in plant science?

24

GENE TRANSFER METHODS

The ultimate goal of every genetic engineering experiment can be achieved only when the gene of interest is successfully introduced into the bacteria or cell. DNA is a negatively charged molecule and the membrane is also negatively charged. Hence due to repulsion force of like charges they never come in contact with each other. If there is no contact, then the transfer of the DNA into the cell becomes impossible. This negative charge repulsion can be removed by using chemicals or by using some physical forces such as high voltage.

ELECTROPORATION

Electroporation is a gene transfer method in which DNA is transferred into cells by using high voltage current for a fraction of second. This method can be applied to transfer DNA into plant cells, animal cells, yeast and bacteria. In this method, cells are placed between the electrodes present in an electroporation chamber along with an ionic solution containing the vector DNA.

An electric pulse generated by a capacitor is applied for about 10–45 milliseconds, which induces pores in the plasma membranes. These pores appear to be round and are present for several minutes after the pulse during which vector DNA passes through the pores into the cell.

The transformation efficiency by this method depends upon the field strength or amount of current applied and the time of application. High voltage (1.5 kV) can be applied for a short duration of 10 milliseconds or a low voltage of 350 V can be applied for a longer duration of 54 milliseconds. Transformation efficiency can also be increased by using various other methods such as adding 13% PEG solution into the electroporation chamber, linearing the DNA or heat shocking (45°C for 5 min.) prior to impulse.

MICROINJECTION

The technique of microinjection is based on the use of glass micropipettes with a 0.5–10 mm diameter tip to directly transfer macromolecules (vectors) into the cytoplasm or nucleus of a recipient cell which is held by a glass suction pipette. This method is widely used for animal and plant cells. In this method, the recipient cell is held with a suction pipette by using suction pressure. Micropipette filled with DNA solution or vector DNA is brought to the recipient cell. Then an appropriate amount of DNA solution is injected into the recipient cell by applying pressure exerted by syringes or mechanized machines connected to the micropipette.

The whole procedure is carried out viewing under an inverted microscope. If large number of cells have to be transformed, then recipient cells are immobilized on a coverslip with poly L-lysine or embedded in a thin layer of agarose in a grid pattern. Two great advantages of this method are the ability to transform the cell with high frequency especially when small number of recipient cells are available, e.g. egg, and non-requirement of pre-treatment procedure for injection. The only disadvantage with this method is that it is very cumbersome and requires expensive, highly skilled personnel.

REVIEW QUESTION

1. Give an account of the different gene transfer methods.

25

APPLICATION OF RECOMBINANT TECHNOLOGY

Genetics is the fundamental biological science, for without genes there is no life. Thus a full understanding of any biological process can be achieved only when there has been a detailed analysis of gene structure and function. In olden days, for the analysis of genes, researchers depended upon mutations, which were time-consuming and cumbersome and without any reproducibility. These problems have been solved with the advent of gene manipulation technologies, which have provided novel solutions to experimental problems in biology. Now rDNA technology has risen to a level where its impact and presence is felt in every area of biology. Some of these are described in this chapter.

APPLICATION IN MEDICINE

Insulin Production

The first licensed drug produced through genetic engineering was human insulin, an important hormone that regulates sugar metabolism. Insulin is produced by a small number of cells in the pancreas and secreted into the bloodstream. The inability to produce insulin results in diabetes, but daily injections of insulin are sufficient to reverse or at least avoid the effects.

In mammals, insulin is expressed as a single chain prepro-hormone, which is secreted through the plasma membrane. A prepro-hormone contains extra amino acids not present in the mature hormone. The pre-sequence present in the middle of chain is important for the current structure and function. During secretion, these extra amino acids are cleaved from the pre-hormone by cellular proteases to release the mature insulin molecule consisting of two short polypeptide chains A and B, linked by two disulphide bonds. The primary problem in the production was getting insulin assembled into this mature form. This problem was solved by creating a single glycosidase–insulin fusion protein, which could be cleaved in a single step to release mature insulin.

Human Growth Hormone (hGH) Production

Some children inherit a condition called hypopituitary dwarfism. They do not grow normally because their bodies make too little growth hormone. They are destined to be dwarfs unless they are treated with hGH. Growth hormone from other animals would not help. Moreover, early preparations of hGH were contaminated with pyrogens or other contaminants.

Growth hormone is a protein containing 191 amino acids that is produced in the pituitary gland and regulates growth and development. The production of hGH was achieved by constructing a hybrid gene from the natural hGH cDNA and synthetic oligonucleotides that encode the amino terminus of the mature form of the protein. This coding sequence was ligated into a plasmid adjacent to a bacterial promoter. Human growth hormone is produced by the bacteria and then secreted with the concomitant removal of the signal peptide by bacterial proteases. The only difference between the secreted hGH and that produced intracellularly is the presence of an amino-terminal methionine on the intracellular expressed molecule. Because the secreted form lacks this methionine, it is called met-less hGH.

Improved Vaccines

Novel, safer vaccines may be another available produce of cloned genes. Usually, vaccines are made from whole viruses that have been either killed or weakened so that they can no longer cause disease but can still arouse the body's immune defences. The problem is that sometimes these viruses are not killed or are not weakened enough, thus they end up causing the disease and sometimes even death.

The first successful subunit vaccine was produced for hepatitis B virus (HBV), which infects the liver. Initial attempts to produce the HbsAg protein in *E. coli* failed, so researchers turned to yeast. The HbsAg gene was inserted into a high-copy yeast expression vector. By growing the yeast in large fermentors, it was possible to produce 50–100 mg/l of protein which closely resembled the natural viral protein. The yeast protein is now used commercially to vaccinate people against HBV infection.

The most successful vaccination campaign to date using a recombinant vaccine is the elimination of rabies with a vaccinia-based vaccine. This has been administered to the wild population of foxes in a large part of central Europe by giving food consisting of chicken heads spiced with the recombinant virus. The epidemiological effect of vaccination has been most evident in eastern Switzerland.

Complex Human Proteins

Most of the proteins are simple and can be expressed in bacteria and yeast. But proteins of medical interest are considerably more complicated in structure and biologically active proteins have proved difficult to produce in bacteria and yeast. Mammalian cells can be useful but they are tricky and expensive to grow. Thus much of the effort has been devoted to setting up fermentor systems for large-scale culture of mammalian cells.

The first drug to be produced commercially by mammalian cell culture was tissue plasminogen activator or tPA, which is administered to heart attack victims. Tissue plasminogen activator is a protease, an enzyme that cleaves other proteins. Rapid administration of a plasminogen activator after a heart attack dissolves the life-threatening clots that lead to irreversible damage of heart muscle. Tissue plasminogen activator is commercially produced from a mammalian cell line carrying a stable, integrated, highly amplified expression vector. Another protein being produced by mammalian cell culture is factor VIII, a protein required for normal clotting of the blood. The factor VIII cDNA has already been cloned and is commercially available.

Identification of Disease-Causing Microorganism

A disease can be caused by a battery of microorganisms. In conventional diagnostic practice, identification of the causative organisms would require cultivation of samples on a variety of different media in a variety of different ways—microscopy, animal cell culture and immunoassay. These procedures are time-consuming, cumbersome and disparate. This whole set of disparate test procedures can be replaced with a single technology called hybridization. Clearly, hybridization of the test sample with a battery of probes should be more simpler and less time-consuming. Hybridization techniques can be used when the infectious agent is present in very low concentration and when the organism cannot be cultured. An important diagnostic tool is *in situ* hybridization for detecting viral pathogens present in the cell.

Hybridization as a diagnostic tool is not restricted to clinical microbiology. There are many applications in plant pathology, monitoring disease control methods and microbial ecology. The disadvantages of nucleic acid probes are that they can determine the presence of microorganism but cannot tell whether it is dead or alive.

Detection of Genetic Disorders

There are several hundred recognized genetic diseases in man which result from single recessive mutations. In some of these, a protein product that is defective or absent have been identified but in many others the nature of the mutation is unknown. For many of these genetic diseases there is no definite treatment and their prevention is the current strategy in many countries.

An essential prerequisite for prenatal diagnosis is the availability of foetal DNA. Foetal cells can be obtained by amniocentesis but this method is not entirely satisfactory for it cannot be carried out during early pregnancy. An alternative approach is to obtain foetal DNA from biopsies of trophoblastic villi in the first trimester of pregnancy. The DNA amplification is done without purification of the cellular DNA with PCR. Of all genetic diseases, the inherited haemoglobin disorder has been the most extensively studied at the DNA level. In what follows, haemoglobinopathies will be taken as examples. Their antenatal diagnosis by recombinant DNA techniques has served as a prototype for other genetic disorders. Clinically the most important haemoglobinopathies are sickle-cell anaemia and the thalassaemia site in genomic DNA. This can be used as a marker for the presence or absence of the defect. Digesting mutated and normal DNA with the restriction enzyme and performing a Southern-blot hybridization with a cloned β-globin DNA probe can therefore detect the mutation. Such an approach is applicable only to those disorders when there is an alteration in a restriction site or where a major deletion or rearrangement alters the restriction pattern.

Monoclonal Antibodies

Researchers have long dreamed of harnessing the specificity of antibodies for a variety of uses that require the targeting of drugs and other treatments to particular sites in the body. It is this use of antibodies as targeting devices that led to the concept of the "magic bullet", a treatment that could effectively seek and destroy tumour cells and infectious agents wherever they reside.

The major limitation in the therapeutic use of antibodies is producing a useful antibody in large quantities. Initially, researchers screened myelomas, which are antibody-secreting tumours for the production of useful antibodies. But they lacked a means to program a myeloma to produce an antibody to their specifications. This situation changed dramatically with the development of monoclonal antibodies. The steps involved in monoclonal antibody production are shown in Figure 25.1.

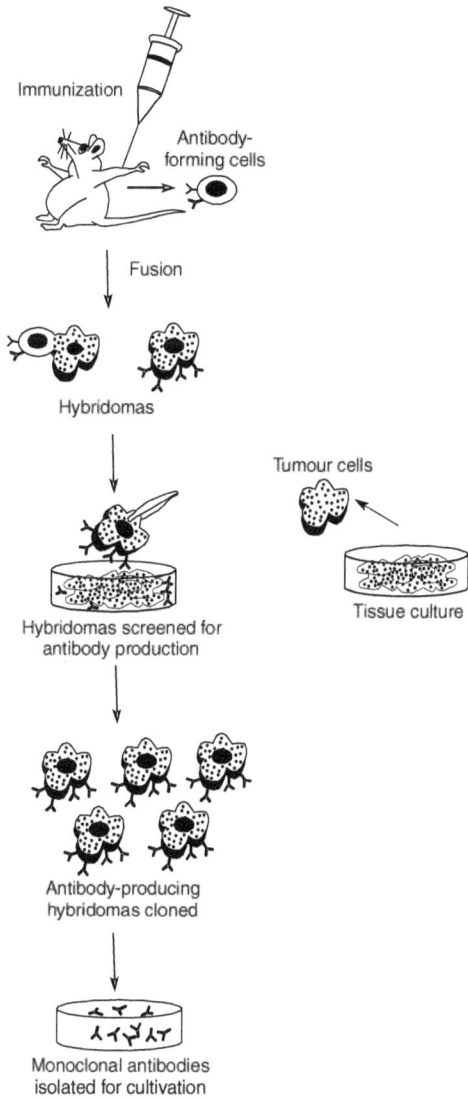

Figure 25.1 Monoclonal antibody production

Protein Engineering

One of the most exciting aspects of recombinant DNA technology is that it permits the design, development and isolation of proteins with improved operating characteristics and even completely novel proteins. The simplest example of protein engineering involves site-directed mutagenesis to alter key residues. The thermostability of T4 lysozyme was increased by 100-fold by the introduction of a disulphide bond.

Many human proteins are being tested as potential therapeutic agents and a number of them are already commercially available. Protein engineering is now being used to generate second-generation variants with improved pharmacokinetics, structure, stability and bioavailability. Insulin is most likely to be assembled as zinc-containing hexamers. This self-association may limit absorption. By making single amino acid substitutions, insulin with high activity and faster absorption were made. Replacing asparagine residue with glutamine altered the glycosylation pattern of tissue plasminogen activator. This in turn significantly increased the circulatory half-life, which in the native enzyme is only 5 min.

APPLICATIONS IN AGRICULTURE

Recombinant DNA technology has not only enhanced the health of humans, but also contributed to exciting developments in agricultural biotechnology. Using rDNA methods, transgenic plants and animals with desirable properties such as resistance to diseases/herbicides have been developed. Flowers with exotic shapes and colours have been genetically engineered by transgenic expression of pigment genes. Recombinant growth hormones are now available for farm animals, resulting in leaner meat, improved milk yield and more efficient feed utilization. In future, transgenic plants and animals may serve as bioreactors for the production of medicinal or protein pharmaceuticals.

Expression of Viral Coat Protein to Resist Infection

Viruses are a serious problem for many agricultural crops and animals. Infections can result in reduced growth, lower yield of products and poor quality. Through a standard genetic trick termed cross-protection, infection of a plant/animal with a strain of virus that produces only mild effect protects the plants/animals against infection by more damaging strains. The same principle/ mechanism when applied for animals, is called as vaccination of monoclonal antibody technology. The procedure for producing monoclonal antibodies, or mABs involves inoculation of the desired antigen into BALB/C mice. After the animal mounts immune response to the antigen, which occurs after 21 days, the animal is sacrificed and its spleen is removed. Spleen cells are separated by mashing the spleen and then fused with myeloma cells. The resulting fused cells or hybridomas retain properties of both parents. They grow continuously and rapidly in cultures like the myeloma cell. Yet they produce antibodies specified by the lymphocytes from the immunized animals. Hundreds of hybridomas can be produced from a single fusion experiment and they are systematically screened to identify those producing large amounts of a desired antibody. Once identified, this antibody is available in limitless quantities.

Expression of Bacterial Toxin

Currently, the major weapons against the attackers are chemical insecticides. But chemicals have impact on the environment. Natural microbial pesticides, such as certain species of *Bacillus thuringiensis* (BT) have been used in a limited fashion for over 38 years. Upon sporulation, these bacteria produce a crystallized protein that is toxic to the larvae of a number of insects. The toxic protein does not harm non-susceptible insects and has no effect on vertebrates. The crystal protein is normally expressed as a large, inactive pro-toxin about 1200 amino acids in length and with 1,20,000 Dalton in weight. The toxin acts by binding to receptors on the surface of midgut cells and blocking the functioning of these cells. Research is on the way for commercial use.

A second approach to the development of insect-resistant plants has been through the transgenic expression of serine protease inhibitors. These proteins are present in a number of plants and function to deter insects by inhibiting serine proteases in the insect digestive system.

Heribicide-tolerant Plants

The presence of weeds in a field of crop reduces the yield. Weed killers or herbicides are not very selective and their current use relies on differential uptake between the weed and the crop plant or on application of the herbicide before planting a field.

With the ability to introduce DNA into plants, researchers are trying to create herbicide-tolerant crops by three strategies.

1. Increasing the level of the target enzyme for a particular herbicide.

2. Expressing a mutant enzyme that is not affected by the compound.

3. Expressing an enzyme that detoxifies the herbicide.

Of the large number of herbicides in use today, only a few of the cellular targets have been characterized. The first strategy is to clone the cellular target genes into the plant so that they are produced in large amounts.

Another strategy for creating herbicide-tolerant plants is the use of mutant forms of bacterial EPSPS enzymes. Genes encoding these mutant enzymes have been cloned from glycophosphate-resistant bacteria and expressed implants. These enzymes have lesser inhibition effect by herbicides. The third strategy for creating herbicide-tolerant plants is by transgenic expression of enzymes that convert the herbicide to a form that is not toxic to the plant. Some plants have developed their own detoxifying system for certain herbicides. But these activities in plants are encoded by a complex set of genes that has not yet been fully characterized.

Pigmentation in Transgenic Plants

Plants are widely used for ornamental purposes so it is not surprising that considerable attempts have been made to develop varieties with flowers with new colours, shape and growth properties by genetic engineering. Although the mechanism of gross protection is not entirely known, it is thought that a particular virus-encoded protein is responsible for the protective effect. Pigmentation in flowers is mainly due to three classes of compounds, the flavonoids, the carotenoids and the betalains. The flavonoids are the best characterized with much information now available concerning their chemistry, biochemistry and molecular genetics.

Experiments are underway to expand the spectrum of colouring of certain floral species by introducing the genes for the entire pigment biosynthesis pathway. A blue rose has never been obtained because rose plants lack the enzymes that synthesize the pigment for blue flower colouration. Although there has been little success, these are exceptions and not the rule. The problem is a phenomenon known as co-suppression. It is a phenomenon where the presence of an extra copy of the gene will suppress the expression of the endogenous genes. An experiment was performed in which a second copy of a petunia pigment gene was introduced into a petunia plant with coloured flowers. It is expected that increased production of the encoded enzyme might produce flowers with a deeper purple colour. But white coloured flowers were produced due to co-suppression. Co-suppression now has been demonstrated in numerous other systems. It does not appear to be a dosage effect resulting from competition for transcription factors. Nor is it a result of a system that detects specific duplicate plant genes. Rather it appears to be the result of a homology-dependent interaction between homologous sequences.

Altering the Food Content of Plants

Starch is the major storage carbohydrate in higher plants. A wide range of different starches is used by the food and other industries. These are obtained by sourcing the starch from different plant varieties coupled with new enzymatic or chemical approach to creating novel starches with new functional properties. The steps involved in the production of transgenic plants are shown in Figure 25.2.

Higher plants produce over 200 kinds of fatty acids some of which have food value. However, many are likely to have industrial (non-food) uses of higher value than edible fatty acids. These are widely used in detergent synthesis. Phytate is the main storage form of phosphorus in many plant seeds but bound in this form it is a poor nutrient for monogastric lower phytate content and higher phosphorus content. Supplementation of broiler diets with transgenic seeds resulted in improved growth rate comparable to diets supplemented with phosphate or fungal phytate.

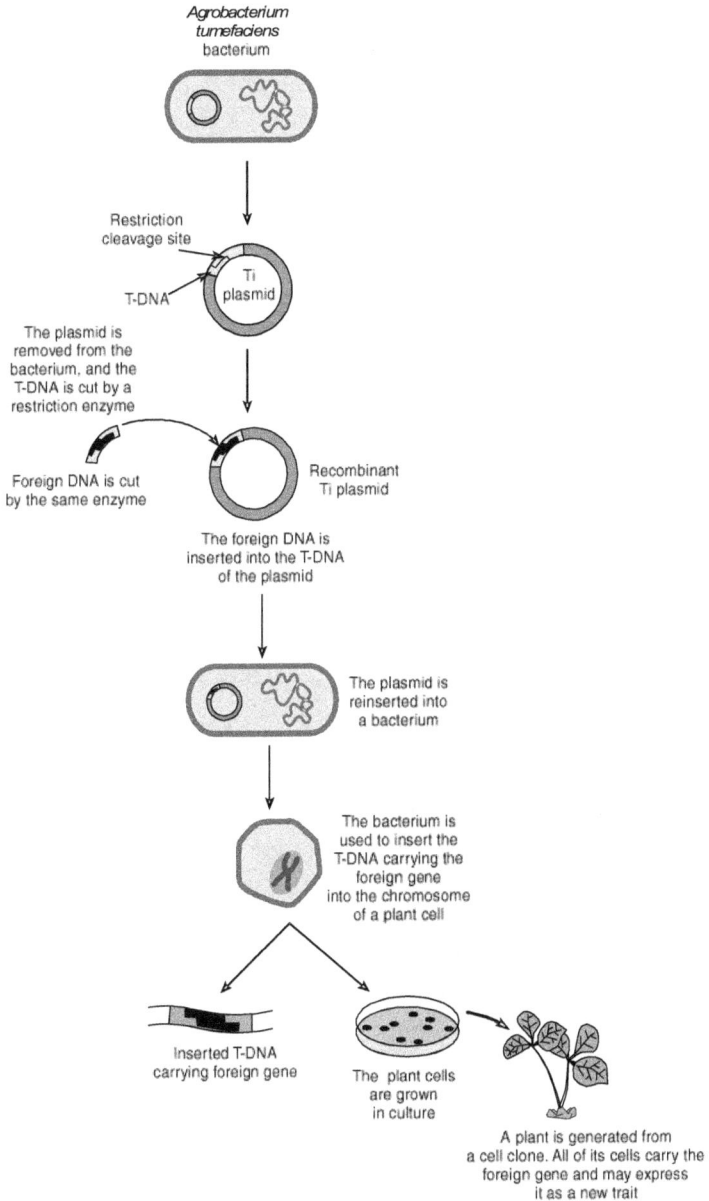

Figure 25.2 Steps for production of transgenic plant

Production of Transgenic Animals

Many generations of selective mating are required to improve livestock and other domesticated animals genetically for traits such as milk yield, wool characteristics, rate of weight gain, etc. The selective mating procedure is time-consuming and costly. Until recently, selective breeding was the only way to enhance the genetic features of domesticated animals. The process of gene manipulation to permanently modify the germ cells of animals is called as "transgenesis" and the animal is called as "transgenic" animal. However, transgenic mice which carry genetic lesions identical to those existing in certain human inherited diseases have been produced. Such mice can be used as models for the development and evaluation of new pharmaceutical entities. The power of this technique can be illustrated by its application to studies on tumour development. In addition to their use as animal models of human diseases, transgenic mice can be used for mutagenicity testing. For many years mice have been used in long-term toxicity testing of new chemicals. In these tests, the animal is acutely or chronically exposed to the test compound and observed constantly for occurrence of tumours.

The most obvious uses for transgenic animals in food production are through direct manipulation of output either by genetically enhancing existing traits or by programming animals to produce novel products. Fortunately, significant progress has been made in producing pharmaceutical proteins in the milk of transgenic farm animals. Apart from the greater regulatory acceptability of farm animals they offer the potential of a high volumetric productivity since milk contains tens of grams of proteins per litre. Milk contains very high level of protein such as casein, β-lactoglobulin and whey acidic proteins. The regulatory sequences from the genes of milk-specific proteins have been cloned and used to control the expressions of heterologous genes in transgenic animals.

One of the goals of transgenesis of dairy cattle is to change the constituents of milk. The amount of cheese produced from milk is directly proportional to the K-casein content. Increasing K-casein production with an over pressed K-casein transgenic is a reasonable likelihood. For a different end use, expression of a lactose transgene in the mammary gland could result in milk that is free of lactose.

Transgenic animals may some day serve as bioreactors that continually secrete high levels of desired proteins into their milk. The proteins would be harvested simply by milking the animals and then standard chromatographic procedures would be used to purify it. The potential applications of transgenic animals in medicine have been expanded with the expression of proteins in blood, tissues and animal organs for transplantation to humans.

INDUSTRIAL APPLICATIONS

Recombinant DNA technology does not just offer novel methods for the generation of proteins, it also provides new ways of making low molecular weight compounds. Good examples, are the microbial synthesis of the blue dye indigo and the black pigment melanin. Neither of these compounds is normally produced by microbes. The cloning of a single gene from *Pseudomonas putida*, that encodes naphthalene deoxygenase, results in the generation of an *E. coli* strain able to synthesize indigo in a medium containing tryptophan. Just as novel proteins can be produced by recombinant DNA techniques so too can novel small molecules. The *Streptomyces coelicolor* gene cluster encoding the biosynthesis of the isochromane-quinone antibiotic was cloned into a variety of other *Streptomyces* sp. producing different isochromane quinines. At least three new antibiotics were detected. Clearly, actinoradin is a novel metabolite in the other *Streptomyces* spp. and is subject to further modifications. Other novel antibiotics produced in this way include 2-norethyromycins A, B, C and D and isovaleyl spiramycin.

GLOSSARY

5′-Phosphate end The phosphate group that is attached to the 5′-carbon atom of sugar. Phosphate group to the 3′-carbon of one nucleotide and the 5′-carbon of another nucleotide.

A (aminoacyl) site The site on the ribosome occupied by an aminoacyl-tRNA just prior to peptide bond formation.

Acentric fragment A chromosomal piece without centromere.

Acrocentric chromosome A chromosome whose centromere lies very near one end.

Activator A eukaryotic specific transcription factor that binds to an enhancer, often far upstream of a promoter.

Adapter A synthetic double-stranded oligonucleotide that is blunt-ended at one end and at the other has a nucleotide extension that can base pair with cohesive end.

A-DNA The form of DNA with high water content.

Affinity tag A short sequence of amino acids that is engineered as part of a recombinant protein and binds to a specific element; compound which facilitates the identification or purification of the recombinant protein.

Albinism The absence of pigment in skin, eyes and hair.

Allele The alternative form of a gene.

Allelic exclusion A process whereby only one Ig light chain and one heavy chain gene are transcribed in any one cell.

Allopolyploidy The polyploidy produced by the hybridization of two species.

Allosteric protein A protein whose shape is changed when it binds a particular molecule. In the new shape, the protein's ability to react to a second molecule is altered.

Allozymes The forms of an enzyme, controlled by alleles of the same locus that differ in electrophoretic mobility.

Alternative splicing   The various ways of splicing out introns in eukaryotic pre-messenger RNAs, so that one gene produces several different mRNAs and protein products.

Ames test   A test for mutagens developed by Bruce Ames which involves the rate of reversion of auxotrophic bacterial strain to prototroph upon exposure to a chemical.

Aminoacyl-tRNA   A charged tRNA; a tRNA with its specific amino acid attached to its 3´-end.

Aminoacyl-tRNA synthetase Enzymes that attach amino acids to their proper tRNAs.

Amino tautomer   The normal tautomer of adenine or cytosine found in nucleic acid.

Amino terminus   The end of a polypeptide with a free amino group; the end at which protein synthesis begins.

Amniocentesis   A procedure for obtaining amniotic fluid from a pregnant women. The chemical content/chromosome abnormalities are observed.

Amorph   A mutation that obliterates gene function.

Anabolic metabolism   The process of building up substance from relatively simple precursors. The *trp* operon encodes anabolic enzymes that build the amino acid tryptophan.

Ancillary protein factors Proteins that are needed for initiation of transcription. These proteins recognize the sequence of DNA close to or overlap with the sequence bound by RNA polymerase itself.

Ancillary site   The sequence present between −30 and −40 position in prokaryotic or eukaryotic gene.

Aneuploids   The individuals or cells exhibiting aneuploidy.

Aneuploidy   The condition of a cell or an organism that has additions or deletions of whole chromosome.

Annealing of DNA   The process of bringing back together the two separate strands of denatured DNA to re-form a double helix.

Antibody   A protein produced by a B-lymphocyte that protects the organism against antigens.

Anti-coding strand   The DNA that forms the template for both the transcribed messenger RNA and the coding strand.

Anticodon   The three-base sequence on transfer RNA complementary to a codon on mRNA.

Anticodon loop The loop, conventionally drawn at the bottom of the tRNA molecule, that contains the anticodon.

Antigen A foreign substance capable of triggering an immune response in an organism.

Antileader A nucleotide sequence in DNA upstream of the initiation codon of a gene which is transcribed but not translated. Antileader generally contains Shine–Dalgarno sequence, which helps in binding of mRNA to the ribosome.

Antimutagen An agent that has the ability to suppress mutagenicity of a mutagen.

Anti-mutator mutation The mutations of DNA pol that decreases the overall mutation rate of a cell or an organism.

Antiparallel strands The strands that run in opposite directions with respect to their 3´ and 5´ ends.

Antisense DNA The sequence of chromosomal DNA that is transcribed.

Antisense RNA The RNA product on *mic* (mRNA-interfering complementary RNA) gene that regulates another gene by base pairing with and thus blocking its mRNA.

Antisense therapy The *in vivo* treatment of a genetic disease by blocking translation of a protein with a DNA or an RNA sequence that is complementary to a specific mRNA.

Anti-sigma factor A protein that interferes with the action of a sigma factor.

Anti-terminator protein A protein that when bound at its normal attachment site, lets RNA polymerase read through normal terminator sequence.

AP endonucleases The endo-nucleases that initiate excision repair at purinic and pyramidinic sites on DNA.

Apoptosis The condition referring to programmed cell death.

Aporepressor A repressor in an inactive form, without its co-repressor.

Aptamer A synthetic nucleic acid that binds to a protein that normally does not bind to nucleic acid.

Apurinic site (AP site) A deoxyribose in a DNA strand that has lost its purine base.

Ascospores The haploid spores found in the asci of ascomycetes fungi.

Ascus   The sac in ascomycetous fungi that holds the ascospores.

Asymmetric transcription   The transcription of only one strand of a given region of a double-stranded polynucleotide.

ATP   Adenosine triphosphate, an energy-rich compound.

ATPase   An enzyme that cleaves ATP, releasing energy for other cellular activities.

Attenuated vaccine   A virulent organism that has been modified to produce a less virulent form, but nevertheless retains the ability to elicit antibodies.

Attenuator   A nucleotide sequence in the 5′-region of a prokaryotic gene that causes premature termination of transcription.

Attenuator region   A control region at the promoter end of repressible amino acid operons, which exerts transcriptional control based on the translation of a small leader peptide.

Authentic protein   A recombinant protein that has all the properties including post-translational modifications.

Autonomous   A term applied to any biological unit that functions on its own, without the help of another unit.

Autonomously replicating sequence (ARS)   A eukaryotic site of the initiation of DNA replication consisting of a 11 base pair consensus sequence.

Autopolyploidy   The term relating to polyploidy in which all the chromosomes come from the same species.

Autoradiography   A technique in which radioactive molecules make their locations known by exposing photographic plates.

Autosomes   The non-sex chromosomes or body chromosomes.

Auxotrophs   The organisms that have specific nutritional requirements.

BAC   Bacterial artificial chromosome.

Back mutation   A mutation that reverses the effect of a previous mutation by restoring the original nucleotide sequence.

Bacmid   A shuttled vector based on the ACMNPV genome that can be propagated in both *E. coli* and insect cells.

Bacterial lawn   A continuous cover of bacteria on the surface of a growth medium.

Bacteriophage   A virus that infects bacteria.

Baculovirus A virus that infects insects.

Base analogue A chemical compound that is structurally similar to one of the bases in DNA and which may act as a mutagen.

Base flipping A process whereby enzymes give access to bases within the DNA double helix by first flipping the bases out of the interior to the outside.

Base pair The complementary nucleotides in DNA are termed as base pair.

Base pair substitution The permanent replacement in chromosomal DNA of a nucleotide pair with another nucleotide pair.

B-Cells The lymphocytes that produce antibodies and are derived from bone marrow.

B-DNA The right-handed, double-helical form of DNA described by Watson and Crick.

β-galactosidase The enzyme that splits lactose into glucose and galactose and coded by a gene *Z* in *lac* operon.

β-galactosidase acetyl transferase An enzyme that is involved in lactose metabolism and is coded by a gene *Z* in *lac* operon.

β-galactoside permease An enzyme that is involved in the *lac* operon.

Bidirectional DNA replication The replication that occurs in both directions at the same time from a common starting point or origin of replication.

Bio-antimutagens These are the suppressors of mutagenicity of mutagens by interfering with the process of mutagenesis.

Biolistics The delivery of DNA to plant and animal cells and organelles by means of DNA-coated pellets that are fired under pressure at high speed.

Bioreactor A vessel in which cells, cell extracts or enzymes carry out a biological reaction.

Biotin labelling The incorporation of biotin-containing nucleotide into a DNA molecule.

Blotting The transfer of a macromolecule by capillary action from a gel to a membrane.

Blunt end The end of a DNA duplex molecule in which neither strand extend beyond the other.

Blunt-end cut To cleave phosphodiester bonds in the backbone of duplex DNA between the corresponding nucleotide pairs on opposite strands.

Blunt-end ligation The process of joining of the nucleotides that are at the ends of two DNA duplex molecules, neither of which has an extension.

Britten–Davidson model A model that explains gene regulation in eukaryotes.

Calibrator lane The lane of a gel that contains electrophoretically separated size markers.

Cap A methylated guanosine added to the 5′ end of eukaryotic mRNA.

Cap binding protein (CBP) A protein that associates with the cap on a eukaryotic mRNA and allows the mRNA to bind to a ribosome.

Capsid The protein shell of a virus.

Carboxyl terminus The end of a polypeptide with a free carboxyl group.

Catabolic metabolism The process of breaking down substances into simpler components. The *lac* operon encodes catabolic enzymes that break down lactose into its component parts.

Catabolite activator protein (CAP) A protein that, when bound with cyclic AMP can attach to sites on sugar-metabolizing operons to enhance transcription of these operons.

Catabolite repression The repression of certain sugar-metabolizing operons in favour of glucose utilization when glucose is present in the environment.

CAT box A conserved sequence found within the promoter region of the protein-encoding genes of many eukaryotic organisms.

cDNA (complementary DNA) The DNA synthesized by reverse transcriptase using RNA as a template.

cDNA clone A double-stranded DNA molecule that is carried in a vector and synthesized *in vitro* from an mRNA sequence by using reverse transcriptase and DNA polymerase.

cDNA library A collection of cDNA clones.

Cell-free system A mixture of cytoplasmic components from cells, lacking nucleic acids and membrane.

Cell line A cell lineage that can be maintained in culture.

Central dogma The original postulate of the way that information can be transferred between DNA and protein.

Centric fragment A piece of chromosome containing a centromere.

Centromere Constrictions in eukaryotic chromosomes on which the kinetochore lies.

Chargaff's rule An observation that states that in the base composition of DNA, the quantity of adenine equals the quantity of thymine, and the quantity of guanine equals the quantity of cytosine.

Charon phages A set of cloning vectors based on a phage.

Chimera It refers usually to a plant or animal that has populations of cells with different genotypes.

Chimeric gene A recombinant gene having regulatory sequence from one gene and coding sequence from another.

Chimeric plasmid The hybrid or genetically mixed plasmid used in DNA cloning.

Chloroplast The organelle that carries out photosynthesis and starch grain formation.

Chromogenic substrate A compound or substance that contains a colour forming group.

Chromosome jumping A technique for isolating clones from a genomic library that are not contiguous but skip regions between known points on the chromosome.

Chromosome walking A technique for studying DNA segments larger than can be individually cloned by using overlapping cloned DNA.

CIB method A technique devised by Muller to screen X-chromosome lethal mutations.

Cis The meaning of the word in "on the nearside off".

Cis-acting A term that describes a genetic element, such as an enhancer, a promoter or an operator that must be on the same chromosome in order to influence a gene activity.

Cis-dominant The mutants that control the functioning of genes on the same piece of DNA.

Cis-trans complementation test A mating test to determine whether two mutants on opposite chromosomes complement each other.

Cistron The term coined by Benzer for the smallest genetic unit that exhibits the *cis–trans* position effect.

Clastogen The agent that causes DNA strands to break.

Cloning site A location on a cloning vector into which DNA can be inserted.

Closed promoter complex The complex formed by relatively loose binding between RNA pol and prokaryotic promoter.

Coding strand The DNA strand with the same sequence as the transcribed mRNA.

Codominance The relationship of alleles in a heterozygote that shows the individual expression of each allele in the phenotype.

Codons The sequences of three RNA or DNA nucleotides that specify either an amino acid or termination of translation.

Cofactor A low molecular weight compound that is required for enzymatic reaction.

Cohesive ends Complementary single-stranded extensions on the ends of duplex DNA molecules.

Colony hybridization A procedure for selecting a bacterial clone containing a gene of interest. DNAs from a large number of clones are simultaneously tested with a labelled probe for gene of interest.

Combinatorial control The transcriptional control in eukaryotes, which involves a large number of polypeptides, many of which recognize specific DNA sequence.

Competence factor A surface protein that binds extracellular

DNA and enables the bacterial cell to be transformed.

Complementarity The correspondence of DNA bases in the double helix so that adenine in one strand is opposite thymine in the other strand.

Complementary RNA The synthetic RNA produced by transcription from a specific DNA single-stranded template.

Complementation The production of the wild type phenotype by a cell or an organism that contains two mutant genes.

Complementation group A group of mutants that cannot complement each other, thus demonstrating that their mutations all lie in the same gene.

Complex gene The gene or its protein product which undergoes various rearrangements, cleavages or modifications before a functional product is obtained.

Composite transposon A transposon constructed of two IS elements flanking a control region that frequently contains host genes.

Compound gene The genes in which coding sequences are separated by non-coding sequences, also called as split gene.

Concatemers The DNA of multiple genome length are called concatemers.

Conditional-lethal mutants A mutant that is lethal under one condition but not under another.

Conformation The shape of a molecule or any other object.

Conjugation A process whereby two cells come in contact and exchange genetic material and occurs only in prokaryotes.

Conjugative plasmid The bacterial plasmids which can be used to transfer genes to bacteria outside their own species.

Consensus sequence A sequence of the common nucleotides found in many different DNA or RNA samples of homologous region.

Conservative replication A postulated mode of DNA replication in which an infant double helix acts as a template for a new double helix.

Conservative transposition The process in which both strands of the transposon DNA are conserved as they leave their original location and move to the new site.

Conserved sequence A sequence found in many different DNA or RNA samples that is invariant in the sample.

Constitutive mutation A mutation that causes a gene to be expressed at all times, regardless of normal controls.

Contigs The genomic libraries of overlapping contiguous clones that cover complete regions of a chromosome.

Copy number The number of molecules of a plasmid contained in a single cell.

Core element An element of the eukaryotic promoter recognized by RNA pol I, which includes the bases surrounding the transcription start site.

Co-repressor The metabolite that when bound to the repressor forms a functional unit that can bind to its operator and block transcription.

Cos The cohesive ends of the linear phage DNA.

Cosmid A hybrid plasmid that contains *cos* site at each end.

Co-transduction The simultaneous transduction of two or more genes.

Crossing over A process in which homologous chromosomes exchange parts by a breakage and reunion process.

C-terminus The last amino acid of a protein.

Cyclic AMP A form of AMP used frequently as a second messenger in eukaryotic hormone action.

Degeneracy Refers to the genetic code and the fact that most amino acids are coded for by more than one triplet codon.

Deletion The loss of chromosomal/base pair segment.

Deletion mapping The mapping mutation by use of overlapping deletion mutants to determine whether or not a mutation includes the site of a mutant gene.

Denaturation The separation of duplex nucleic acid molecules into single strands.

Denatured DNA The duplex DNA that has been converted into single strands by breaking hydrogen bonds of complementary nucleotide pairs.

Density-gradient centrifugation A method of separating molecular entities by their differential sedimentation in a centrifugal gradient.

Derepression The displacement of a repressor protein from a promoter region of DNA.

Desmutagen An agent which suppresses mutagenic activity of a mutagen by directly inactivating the mutagen by destroying the oxygen radicals produced.

Dicentric chromosome A chromosome with two centromeres.

Dideoxynucleotide A modified nucleotide that lacks the 3′-hydroxyl group.

Dimorphic gene A gene whose protein product exists in two forms—inactive and active.

Direct DNA repair The DNA repair that involves neither removal nor replacement of bases nor nucleotides.

Direct repeats The term which refers to two or more identical nucleotide sequences present in a single polynucleotide.

DNA clone A section of DNA that has been inserted into a vector molecule.

DNA construct A cloning vector with a DNA insert.

DNA cross-linking The process in which interstrand thymines of DNA form dimers thus blocking replication.

DNA fingerprint A set of DNA fragments that are characteristic for a particular source of DNA such as an insert of a clone.

DNA fingerprinting A comparative diagnostic technique that characterizes the DNA of an organism or a sample.

DNA glycosylases Endonucleases that initiate excision repair at the sites of various damaged or improper bases.

DNA gyrase A type-II DNA topoisomerase.

DNA hybridization The pairing of two DNA molecules, often from different sources, by hydrogen bonding between complementary nucleotides.

DNA ligase An enzyme that repairs single-stranded discontinuities in double-stranded DNA molecules.

DNA polymerase An enzyme that links an incoming deoxyribonucleotide, which is determined by complementarity to deoxyribonucleotide in a template DNA strand, with a phosphodiester bond to the 3′-hydroxyl group of the last incorporated nucleotide of the growing strand during replication.

DNA probe A segment of DNA that is labelled so that, after a DNA hybridization reaction, any base pairing between the probe and complementary base sequence in a DNA sample can be detected.

DNA repair The correction of nucleotide errors introduced during DNA replication or by mutagens.

DNA sequencing The determination of the order of nucleotides in a DNA molecule.

DNA strand One of the two DNA strands in a double helix.

DNA topoisomerase An enzyme that introduces or removes turns from the double helix by transient breakage of one or both polynucleotides.

DNase An enzyme that degrades DNA.

Dominant The condition in which a heterozygous and homozygous genotype determine the same phenotype.

Dot blotting A blotting technique of DNA already cloned that eliminates the electrophoretic separation step.

Double crossover The two simultaneous reciprocal breakage and reunion events taking place between two DNA molecules.

Double digest The product formed when two different restriction endonucleases act on the same segment of DNA.

Downstream The process taking place towards the 3′-end of a polynucleotide.

Duplicate genes Either one or both dominant genes together producing the same phenotype to give 15:1 ratio.

Duplication The occurrence of a segment more than once in the same chromosome or genome.

Dyad The two sisters chromatids attached to the same centromere.

Editing The process of checking each nucleotide for complementarity with its base-pairing partners.

Effector gene The gene which drives transcription of another gene with the help of a promoter from another gene.

Electrophoresis The movement charged of suspended particles in solution in an electric field.

Electroporation A process whereby cell membranes are made permeable to DNA by applying an intense electric current.

ELISA Enzyme linked immunosorbent assay.

Elongation complex The form of RNA pol II that actively carries out basal transcription.

Elongation factors The proteins necessary for the proper elongation and translocation processes during translation.

End labelling The addition of radioactively labelled group to one end (5´ or 3´) of a DNA strand.

Endogenote The part of the bacterial chromosome that is homologous to a genome fragment transferred from the donor to the recipient cell in the formation of a merozygote.

Endonuclease An enzyme that breaks strands of DNA at internal positions, some of which are involved in recombination of DNA.

Enhancer A substance or object that increases a chemical activity or a physiological process or a DNA sequence that influences transcription of a nearby gene.

Episome A genetic element that may be present or absent in different cells and that may be inserted in a chromosome or independent in the cytoplasm.

Epistasis The interaction between products of non-allelic genes. The genes which are suppressed are said to be hypostatic.

Error-prone repair A mechanism by *E. coli* cells to replicate DNA that contains thymine dimers.

Excision repair A process whereby cells remove part of a damaged DNA strand and replace it through DNA synthesis using the undamaged strand as a template.

Exogenote Chromosomal fragments homologous to an endogenote and donated to a merozygote.

Exons The segments of a eukaryotic gene that correspond to the sequences in the final processed RNA transcript of that gene.

Exon shuffling The hypothesis put forward by Walter Albert that exons, code for the functional units of a protein and that the evolution of new genes proceeds by recombination.

Exonuclease An enzyme that digests DNA or RNA beginning at the ends of strands.

Expression vector A plasmid or phage cloning vehicle specifically constructed so as to achieve efficient transcription of the cloned DNA fragment and translation of its mRNA.

F-duction The transfer of bacterial genes from one cell to another on an F⁻ plasmid.

Feedback inhibition A post-translational control mechanism in which the end product of a biochemical pathway inhibits the activity of the enzyme of the same pathway.

Fertility factor The plasmid that allows a prokaryote to engage in conjugation with and pass DNA into F⁻ cell.

Fluorescent *in situ* hybridization A technique in which a fluorescent dye is attached to a nucleotide probe that then binds to a specific site on a chromosome and makes itself visible by its fluorescence.

Footprinting A technique to determine the length of nucleic acid in contact with a protein.

F-Pili Sex pili.

Frameshift A mutation in which there is an addition or deletion of nucleotides that causes the codon reading frame to shift.

Fusion protein The product of two or more coding sequences from different genes that have been cloned together and which, after translation, form a single polypeptide sequence. It is also called as chimeric protein.

Galactoside permease An enzyme encoded in the *E. coli lac* operon that transports lactose into the cell.

G-cap The 5′-terminal methylated guanine nucleotide that is present on many eukaryotic mRNA.

Gene The basic unit of heredity.

Gene bank A population of organisms each of which carries a DNA molecule that is inserted into a cloning vector.

Gene cloning The method of generating many copies of a gene by inserting it into an organism, where it can replicate along with the host.

Gene concept The method of understanding a gene in terms of its structure and function.

Gene expression The process by which gene products are made.

Gene library A large collection of cloning vectors containing a complete set of fragments of the genome of an organism.

Gene map The linear array of genes of a chromosome.

Gene mutation A mutation confined to a single gene.

Gene pool The sum total of all different alleles in the breeding members of a population.

Genetic code The set of 64 codons and the amino acids they stand for.

Genetic mapping Determining the linear order of genes and the distance between them.

Genetic screening The testing of individuals for a particular trait.

Gene transfer The passage of a gene or group of genes from a donor to recipient organisms.

Genome One complete set of genetic information from a genetic system.

Genome organization The arrangement of the genetic material in the cells of an organism.

Genomic library A set of clones containing DNA fragments derived directly from a genome.

Glycosidic bond The bond linking the base to the sugar in RNA or DNA.

Glycosylation The covalent addition of sugar to protein.

Group-I introns Self-splicing introns in which splicing is initiated by a free guanosine.

Gyrase An enzyme that relaxes supercoiling caused by unwinding of double helix of DNA.

Hairpin loop A double-helical region formed by base pairing between adjacent complementary sequences in a single strand of RNA or DNA.

Haploid An organism or cell having only one complete set of chromosomes.

Hapten A small molecule that acts as an antigen when conjugated to a protein.

Helicase A protein that unwinds DNA.

Helix Any structure with a spiral shape.

Helper plasmid A plasmid that provides a function to another plasmid in the same cell.

Helper virus A virus that provides a function to another virus or viral genome.

Heterochromatin A chromatin staining darkly even during interphase often containing repetitive DNA with few genes.

Heterogeneous nuclear mRNA The original RNA transcripts found in eukaryotic nuclei before post-transcriptional modifications.

Heterologous probe A DNA probe that is derived from one organism and is used to screen for a similar DNA sequence in a clone derived from another organism.

Heteromer A protein with two identical polypeptide chains.

Heteromorphic The chromosomes that are not morphologically identical.

Hfr High frequency recombination strain.

Histones A group of proteins rich in basic amino acids.

Holliday junction A junction point between two cross-linked DNA double helices. It is an intermediate stage in DNA recombination.

Homologous chromosomes The chromosomes that occur in pairs and are generally similar in size and shape.

Homopolymer A nucleic acid strand that is composed of one kind of nucleotide.

Host A microorganism or cell that maintains a cloning vector.

Hybrid gene The combination of two genes or the parts of two genes in the correct reading frames that encodes a single protein that has amino acid sequences from both genes.

Hybrid plasmid A plasmid that contains an inserted piece of foreign DNA.

Hybridization The pairing of two polynucleotide strands, often from different sources by hydrogen bonding between complementary nucleotides.

Hybridoma The product of the fusion of a myeloma cell with an antibody-producing lymphocyte.

Idiogram A photograph or diagram of the chromosomes of a cell arranged in an orderly fashion.

Immunoassay A protocol that uses antibody specificity to detect the presence of a particular compound in a biological sample.

Incision Nicking a DNA strand with an endonuclease.

Incompatibility of plasmids The inability of certain types of plasmids to co-exist in the same cell.

Induced mutations The genetic changes produced by some physical or chemical agent or under changed growth conditions.

Inducer A molecule that induces expression of a gene or operon by binding to a repressor protein and thereby prevents the repressor from attaching to its operator site.

Induction A chemical or physical treatment which results in excision from the host genome of the integrated form of a lysogenic phage.

Initiation codon The codon, usually but not exclusively 5′ – AUG – 3′, which indicates the point at which translation of an mRNA should begin.

Initiation complex The complex which comprises mRNA, a small ribosomal subunit and aminoacylated initiator tRNA.

Initiation factor Protein molecules that play an ancillary role in the initiation stage of translation.

Initiation site The site on the DNA molecule where the synthesis of RNA begins.

Integrating vector A vector that is designed to integrate cloned DNA into the host cell chromosomal DNA.

Interferons A heterogeneous family of multifunctional cytokines that were originally identified as proteins responsible for the induction of cellular resistance to viral infection.

Internal control region The part of a eukaryotic promoter recognized by RNA polymerase III that lies inside the gene's coding region.

Intron A segment of DNA that is transcribed, but removed from within the transcript by splicing.

Inversion Alteration of the sequence of a DNA molecule by removal of a segment followed by its reinsertion in reverse orientation.

Inverted repeats A symmetrical sequence of DNA, reading the same forward on one strand and backward on the opposite strand.

IS element A short DNA sequence found in bacteria capable of transposition.

Isoschizomers Two or more restriction enzymes that recognize and cut the same restriction site.

Jumping gene The gene which keeps on changing its position in a chromosome and also between the chromosomes in a genome.

Karyotype A pictorial or photographic representation of all the chromosomes.

Keto-tautomer The normal tautomer of uracil, thymine or guanine found in nucleic acid.

Kinase An enzyme which will remove or add a phosphate group to a protein or nucleic acid.

Klenow fragment A fragment of DNA pol I, created by cleaving with a protease, and lacks 5′–3′ exonuclease activity.

Knockout mice Transgenic mice that have been made homozygous for a non-functioning allele at a particular locus.

Label A compound or atom that is either attached to or incorporated into a macromolecule.

***lac* operon** The inducible operon, including three loci involved in the uptake and breakdown of lactose.

Lactose repressor The regulatory protein that controls transcription of the *lac* operon in response to the levels of lactose in the environment.

Lagging strand Strand of DNA being replicated discontinuously.

Leader The length of messenger RNA from the 5′ end to the initiation codon ATG.

Leader peptidase An enzyme used to cleave leader signal from the pre-protein to be secreted in bacterial cells.

Leader peptide In bacteria proteins destined to be secreted are synthesized as pre-proteins with N-terminal signal sequences sometimes termed leader peptides.

Leader peptide gene A small gene within the attenuator control region of a repressible amino acid operon.

Leader strand The strand of DNA being replicated continuously.

Lethal allele An allele that causes mortality.

Ligase An enzyme that joins the ends of two strands of nucleic acid.

Ligase chain reaction A technique for determining the presence or absence of a specific nucleotide pair within a target gene.

Ligation The joining of two DNA molecules by the formation of phosphodiester bonds.

Light repair The direct repair of a thymine dimer by the enzyme DNA photolyase.

LINES (Long interspersed elements) The sequences of DNA up to 7000 bp in length, interspersed in eukaryotic chromosomes in many copies.

Linkage The association of loci on the same chromosome.

Linkage group The association of loci on the same chromosome.

Linker A small segment of DNA that contains a restriction site.

Lipofection The delivery into eukaryotic cells of DNA, RNA or other compounds that have been encapsulated in an artificial phospholipid vesicle.

Locus The position of a gene on a chromosome.

Long terminal repeats A DNA strand that is synthesized during the polymerase chain reaction.

Marker A gene or mutation that serves as a signpost at a known location in the genome.

Messenger RNA A transcript of a protein-encoding gene which acts as a template for translation.

Methylation The addition of a methyl group to a macromolecule; the addition of a methyl group to specific cytosine.

Microinjection The injection of purified gene directly into the nucleus of an animal cell through micropipette.

Microprojectile method A method of gene transfer where gold or tungsten particles are coated with DNA.

Microsatellite A short DNA sequence repeated many times in tandem.

Minus 35 box An *E. coli* promoter element centred about 35 bases upstream from the start of transcription.

Minus ten box Pribnow box.

Mismatch repair The correction of a mismatched base incorporated by accident into a newly synthesized DNA.

Mis-sense mutation A change in a codon that results in an amino acid change in the corresponding protein.

Molecular farming The use of transgenic animals/plants as factories for producing speciality chemicals and pharmaceuticals.

Monocistronic mRNA An RNA molecule mostly in eukar-yotes that contains information from only one cistron.

Monoclonal antibody A specific antibody produced by a hybridoma cell line against one specific antigen determinant.

Multiple cloning site A region in certain cloning vectors, such as the pUC plasmid.

Mutagen A mutation-causing agent.

Mutagenesis A chemical or physical treatment that changes the nucleotides of the DNA of an organism.

Mutant An organism that has suffered at least one mutation.

Mutation An alteration in the nucleotide sequence of a DNA molecule.

Mutation rate The rate of mutational change from one allelic form to another.

Native protein The naturally occurring form of a protein.

Negative control A system of regulation of transcription that requires the removal of a repressor.

Nested primers Two or more sets of pairs of PCR primers that bind to neighbouring DNA sequences.

Nick To break a phosphodiester bond in the backbone of one of the strands of a duplex DNA molecule.

Nonsense codon One of the messenger RNA sequence that signals termination.

Northern blotting Similar to Southern blotting, but RNA is transferred from the gel on to the membrane.

N-terminus The first amino acid of a protein.

Nuclear cloning The production of an organism by placing a nucleus from somatic cells into an enucleated egg.

Nucleolus The site of ribosome construction.

Nucleoprotein The substance of eukaryotic chromosomes consisting of proteins and nucleic acids.

Nucleotide A base that is covalently linked to a five-carbon sugar; a nucleoside linked to phosphate.

Nucleotide excision repair The DNA excision repair mechanism responsible for repairing thymine dimers.

Ochre codon The nonsense codon UAA.

Octad An ascus containing eight ascospores.

Octopine A rare amino acid derivative which is produced by a certain type of crown gall tissue.

Okazaki fragment One of the short segments of RNA-primed DNA that is synthesized during replication.

Oligomeric protein A protein that contains more than one polypeptide subunit.

Oligonucleotide therapy A linear sequence of few nucleotides.

Opal codon The nonsense codon UGA.

Open promoter complex The complex formed between *E. coli* RNA pol and a promoter in which the double helices partially unwind in readiness for the start of RNA synthesis.

Open reading frame A series of codons with an initiation codon at the 5´ end but no termination codon; a DNA sequence, which looks like a gene but to which no function has been assigned.

Operator A nucleotide sequence element to which a repressor protein attaches.

Operon A system of cistrons, operator and promoter sites by which a given genetically controlled metabolic activity is regulated.

Origin of replication The unique spot in a replicon where replication begins.

Origin of transfer When gene transfer occurs through plasmids, one of the two strands of plasmid DNA is nicked at a site called "*ori*T".

Palindromic sequences Complementary DNA sequences that are the same when each strand is read in the same direction.

Partial digest Treatment of a DNA sample with a type-II restriction endonuclease.

pBR 322 A particular type of artificial plasmid.

PCR (Polymerase chain reaction) A technique of amplifying a specific segment of DNA.

Pedigree A diagrammatic representation of the history of a trait in a multigeneration family.

Peptide A short chain of amino acids.

Peptide bond The covalent bond between the free carboxyl group of the α-carbon of one amino acid and the free amino group of the α-carbon of an adjacent amino acid in a peptide.

Peptide vaccine A short chain of a.a. that can induce antibodies.

Peptidyl site The portion of a ribosome where the tRNA with the peptide chain participates in peptide bond formation.

Peptidyl transferase The enzyme activity responsible for peptide bond synthesis during translation.

Peptidyl-tRNA The tRNA that has a growing peptide chain attached to it during translation.

Phosphorothioate linkage The linkage between nucleotides after a sulphur group replaces an available oxygen of a phospho-diester linkage.

Photoreactivation A light-induced reversal of ultraviolet light causing injury to cells.

Plaque A clear area that is visible in a bacterial lawn on an agar plate.

Plasmid An autonomous, self-replicating extrachromosomal DNA molecule.

Poly(A) polymerase The enzyme responsible for polyadenylation of a eukaryotic mRNA molecule.

Poly(A) tail The initially long sequence of adenine nucleotides at the 3′ end of mRNA.

Poly(A) T technique A method of inserting foreign DNA into a vehicle by making 5′-poly (A) and 5′-poly(dT) tails on the vehicle and foreign DNA.

Polyadenylation The addition of adenine residues to the 3′ – end of eukaryotic mRNA.

Polyadenylation signal A sequence that terminates transcription and provides a recognition site at the end of an mRNA for the enzymatic addition of adenine residues.

Polycistronic RNA An mRNA that encodes two or more proteins.

Polyhedron A structure in which baculovirus genomes are inside a cylindrical nucleocapsid, which is enclosed in an occlusion body composed of the protein polyhedrin.

Polylinker A synthetic DNA sequence that contains a number of different restriction endonuclease site.

Polymer A macromolecule made up of a series of covalently linked monomer.

Polynucleotide A linear series of 20 or more nucleotides linked by phosphodiester bonds.

Post-replicative repair A DNA repair process initiated when DNA polymerase bypasses a damaged area.

Post-translational modification The specific addition of phosphate group sugars or other molecules to a protein after it has been synthesized.

Pribnow box Relatively invariable six-nucleotide DNA sequence TATAAT, located in eukaryotic promoters.

Primary transcript The unprocessed RNA that is transcribed from a eukaryotic structural gene that has exons and introns.

Primase The RNA polymerase enzyme that synthesizes the primer needed to initiate replication of a DNA polynucleotide.

Primer A short oligonucleotide that hybridizes with a template strand and provides a 3′ – OH end for the initiation of nucleic acid synthesis.

Probe A DNA sequence that is used to detect the presence of a complementary sequence by hybridization.

Progeny The offspring of a mating.

Promoter A segment of DNA to which RNA polymerase attaches. It usually lies upstream of 5´ end of a gene.

Proofreading The ability of a DNA pol to correct misincorporated nucleotides as a result of its 3´ → 5´ exonuclease activity.

Quenching Quickly chilling heat-denatured DNA to keep it denatured.

Rapid-lysis mutants These mutants display a change in the pattern of lysis of *E.coli* at the end of an infection by a T-even phage.

Reading frame The sequence of triplet codons contained in any DNA/RNA sequence beginning at translation initiation codon.

Recognition site A sequence of bases within the promoter region that serves to recognize RNA pol molecule.

Recombinant DNA A DNA molecule created in the test tube by ligating together pieces of DNA that are not normally contiguous.

Recombinational repair Filling a gap in one strand of duplex DNA by retrieving a homologous single strand from another duplex.

Recon The smallest segment of DNA or subunit of a cistron that is capable of recombination.

Regulator gene A gene that codes for a protein such as a repressor.

Release factors Proteins (RF1, RF2, RF3) in prokaryotes responsible for termination of translation and release of the newly synthesized polypeptide.

Relaxed plasmid A plasmid whose replication is not linked to replication of the host genome.

Renaturation The return of a denatured molecule back to its natural state.

Replica plating A technique used to make replicas of bacterial colonies from a master plate to fresh plate.

Replicase An enzyme that unwinds double helical DNA.

Replication The process of copying; replication of DNA.

Replication fork The region of a double-stranded DNA molecule that is being unwound to enable DNA replication to occur.

Replisome A sequentially replicating segment of a nucleic acid controlled by a sub-segment known as a replicator.

Repressible control Repression of reproduction of the enzymes by action of end product.

Repression The ability of a bacteria to prevent synthesis of certain enzymes when their products are present.

Repressor A regulator molecule produced by a regulator gene that can combine with and repress action of an associated operator.

Resolvase An enzyme actively involved in recombination between transposons.

Restriction enzyme Any of a group of enzymes that break internal bond of DNA at highly specific points.

Restriction map The linear array of restriction endonuclease sites on a DNA molecule.

Restriction site The base sequence at which a restriction endonuclease cuts the DNA molecule.

Reverse mutation Any second mutation which converts mutant into wild type.

Reverse transcriptase An enzyme that synthesizes a DNA copy on an RNA template.

Rho-factor A protein that is required for termination of transcription.

Ribosome An RNA molecule that has catalytic activity.

Ribosome binding site The nucleotide sequence that acts as the site for attachment of a ribosome to an mRNA molecule.

R-Loop The structure formed when an RNA strand hybridizes with its complementary strand in a DNA duplex, thereby displacing the original strand of DNA in the form of a loop.

RNA polymerase An enzyme capable of synthesizing an RNA copy of a DNA template.

RNase An enzyme that hydrolyses RNA.

RNA splicing The removal of large non-coding introns from the primary transcript.

S1-nuclease An enzyme that specifically degrades single-stranded DNA.

S-adenosyl methionine A molecule which donates its methyl group to cytosine or adenine during DNA methylation.

scRNA Any one of several small cytoplasmic RNA.

SDS-PAGE Sodium dodecyl sulphate-polyacrylamide gel electrophoresis.

Semiconservative replication The mode of DNA replication in which each daughter double helix comprises one polynucleotide from the parent and one newly synthesized polynucleotide.

Sensitivity For diagnostic tests, the ability of an assay to detect the smallest amount of the target molecule.

Sexduction Incorporation of bacterial chromosomal genes in the fertility plasmid.

Sex factor An episome of bacteria which causes chromosome breakage during conjugation and facilitates genetic exchange.

Shine–Dalgarno   sequence Ribosomal binding site.

Short tandem repeat A DNA sequence with a sequential repeating set of two–four pairs.

Shot gun cloning The cloning of an entire genome in the form of randomly generated fragments.

Shuttle vector A plasmid cloning vehicle, usually a plasmid that can replicate in two different organisms.

Sigma factor An accessory bacterial protein that directs the binding of RNA polymerase to specific promoters.

Silencer The DNA sequence that helps to inhibit transcription, located at a long distance from the core promoter.

Silent mutation An alteration in a DNA sequence which does not affect the expression or functioning of any gene or gene product.

Sine A type of dispersed repetitive DNA.

Site-specific mutagenesis A technique to change one or more specific nucleotides in a cloned gene in order to create an altered form of a protein with a specific amino acid change.

SOS repair system The repair systems induced by the presence of single-stranded DNA that usually occurs from post-replicative gaps caused by various types of DNA damage.

Southern blotting A technique for transferring denatured DNA molecules that have been separated on the gel on to a matrix or nitrocellulose membrane.

Spacer DNA Regions of non-transcribed DNA between transcribed segments.

Specialized   transduction A type of transduction where only a few bacterial genes are transferred because the phage has only specific sites of integration on the host chromosome.

Spliceosome The protein–RNA structure believed to be responsible for splicing.

Split gene A discontinuous gene.

Staggered cot Symmetrically cleaved phosphodiester bonds that lie on both strands of duplex DNA but are not opposite to one another.

Start codon A codon which codes for initiation of protein synthesis.

Start point The position on DNA corresponding to the first base incorporated into RNA.

Sticky end The end of double-stranded DNA molecules where there is a single-stranded extension.

Strand A linear series of nucleotides that are linked to each other by phosphodiester bond.

Tailing The *in vitro* addition of the same nucleotide by the enzymes terminal transferase to the 3′-OH end of a duplex DNA molecule.

Tandem repeats Direct repeats that are adjacent to each other with no intervening DNA.

Target site duplication A sequence of DNA that is duplicated when a transposable element inserts, usually found at each end of the insertion.

TATA box A consensus sequence surrounding the lariat branch point of prokaryotic mRNA.

Telomerase The enzyme that maintains the ends of chromosomes by synthesizing telomeric repeat sequences.

Template A pattern serving as a mechanical guide.

Termination codon One of the three codons that marks the position where translation of an mRNA should stop.

Termination factors (TF) The proteins required to obtain release of newly synthesized polypeptide chains from tRNA.

Terminator sequence A sequence in DNA that signals the termination of transcription to RNA polymerase.

Thymine dimer Hydrogen bonding of two molecules of thymine by action of ultraviolet light.

Tissue plasminogen activator A protein involved in dissolving blood clots.

Topoisomerase An enzyme that can relieve supercoiling in DNA by creating transitory breaks in the helical backbone.

***Trans*-acting** The normal mode of action of most recessive mutations.

Transcript An RNA molecule that has been synthesized from a specific DNA template.

Transcriptional elements The DNA sequence in eukaryotic genes which controls transcription of the gene.

Transcription factors Eukaryotic protein that aids RNA pol to recognize promoters.

Transfection The introduction of foreign DNA into animal cells.

Transformant A cell that has become transformed by the uptake of naked DNA.

Transformation The uptake of DNA and its establishment in bacteria.

Transformation efficiency The number of cells that take up foreign DNA as a function of the amount of added DNA.

Transgenic Eukaryotic organisms that have taken up foreign DNA.

Translation The synthesis of a polypeptide, the amino acid sequence of which is determined by the nucleotide sequence of an mRNA.

Transposase The enzyme that catalyses transposition of a transposable genetic element.

Transposition The process whereby a transposon or insertion sequence inserts into a new site on the same or another DNA molecule.

Transposon A DNA sequence that can insert randomly into a chromosome, exit the site and relocate itself at another chromosome location.

Triplet code A group of three successive nucleotides in RNA or DNA that specifies a particular amino acid.

Underwinding The winding of DNA produced by negative supercoiling.

Unstable gene A gene which mutates frequently.

Upstream Towards the $5'$ end of a polynucleotide.

URF (Unidentified reading frame) An open reading frame recognized from a DNA sequence for which no genetic function is known.

U-snRNPs U-class small ribonucleoproteins.

Vector A DNA molecule capable of replication into which a gene is inserted by recombinant DNA techniques.

Vehicle The host organism used for the replication or expression of a cloned gene.

Virulence The degree of pathogenicity of an organism.

Watson–Crick base pairing The normal base pairing.

Western blotting The process involving transfer of proteins from a gel to membrane.

Wild type A genetic term that denotes the most commonly observed pheno-type.

Wobble base pairing The pairing of mRNA codon with tRNA anticodon.

Yeast artificial chromosome A high-capacity cloning vector that replicates in yeast cells. It contains yeast chromosomes and is able to clone very large pieces.

Yeast centromere plasmid A 2 µm long plasmid without any known function found in many strain of yeast.

Yeast replicative plasmid Plasmids containing ARS derived from yeast chromosome.

Z-DNA The DNA in which sugar and phosphate linkage follow a zig-zag pattern.

Zinc-finger Sequence-specific DNA-binding proteins that contain domains that bind Zn^{++}.

FURTHER READINGS

Alberts Bruce, Johnson Alexander, Lewis Julian, Raff Martin, Roberts Keith and Walter Peter. *Molecular Biology of the Cell*, 4th edn. Garland Publishing, New York. 2002.

Alex Prokop, Rakesh, K. Bajpai and Chester, S. Ho. *Recombinant DNA Technology and Applications*. McGraw-Hill. 1991.

Anthony, J.F., Griffiths Susan, Wessler Richard, R., Lewontin, C., William, M. Gelbart, David, T. Suzuki and Jeffrey, H. Miller. *An Introduction to Genetic Analysis*, 8th edn. W. H. Freeman. 2004.

Anthony, J.F., Griffiths Susan, Richard, R., Lewontin, C., Jeffrey, H. Miller, William, M. Gelbart. *Modern Genetic Analysis : Integrating Genes and Genomes*, 2nd edn. W. H. Freeman. 2002.

Benjamin Lewin. *Genes VIII*. Pearson Prentice Hall. USA. 2004.

Bernard, R. Glick and Jack, J. Pasternak. *Molecular Biotechnology: Principles and Applications of Recombinant DNA*. ASM Press. 2003.

Berg Jeremy, M., Tymoczko John, L. and Stryer Lubert. *Biochemistry*. W.H. Freeman and Co. New York. 2002.

Brown, T.A. *Gene Cloning and DNA Analysis: An Introduction*. Blackwell Scientific Publication. 2001.

Brown, T.A. *Gene Cloning*. IRL Press Oxford Univ. Press. 1998.

Brown, T.A. *Genomes*. IRL Press Oxford Univ. Press. 1999.

Colin Ratledge and Bjorn Kristiansen. *Basic Biotechnology*. Cambridge Univ. Press. 2005.

Daniel, L. Hartl and Elizabeth, W. Jones. *Genetics: Analysis of Genes and Genomes*, 6th edn. Jones & Bartlett Publishers. 2004.

Donald Voet and Judith, G. Voet. *Biochemistry*, 2nd edn. Wiley. 1995.

Geoffrey, M. Cooper and Robert, E. Hausman. *The Cell: A Molecular Approach*, 3rd edn. Sinauer Associates. 2003.

Geoffrey, M. Cooper and Sunderland. *The Cell - A Molecular Approach*, 2nd edn. MA Sinauer Associates Inc. 2000.

Gerald Karp. *Cell and Molecular Biology*, 4th Rev. edn. John Wiley & Sons Inc. 2004.

Griffiths Anthony, J.F., Gelbart William, M., Miller Jeffrey, H., and Lewontin Richard C., *Modern Genetic Analysis*. W. H. Freeman & Co. New York.1999.

Gupta, P.K. *Elements of Biotechnology*. Rastogi Publications. 1998.

James, D. Watson, Tania, A. Baker, Stephen, P. Bell and Alexander Gann. *Molecular Biology of the Gene*, 5th edn. Benjamin Cummings. 2003.

Jeremy, W. Dale and Malcolm von Schantz. *From Genes to Genomes: Concepts and Applications of DNA Technology*. John Wiley & Sons. 2002.

James Greene. *Recombinant DNA Principles Methodologies*. CRC. 1998.

Jeremy, W. Dale and Simon, F. Park. *Molecular Genetics of Bacteria*, 4th edn. John Wiley & Sons. 2004.

Kumar, H.D. *Modern Concepts of Biotechnology*. Vikas Publishing House Pvt. Ltd. 2002.

Lodish Harvey, Berk Arnold, Zipursky, S. Lawrence, Matsudaira Paul, Baltimore David and Darnell James, E. *Molecular Cell Biology*, 4th edn. W.H. Freeman & Co. New York. 1999.

Nancy Jo Trun and Trempy, J.E. *Fundamental Bacterial Genetics*. Blackwell Science. 2003.

Primrose, S.B. and Old, R.W. *Principles of Gene Manipulation*, 5th edn. Blackwell Scientific Publication. 1994.

Richard, J. Reece. *Analysis of Genes and Genomes.* John Wiley & Sons. 2004.

Robert, H. Tamarin. *Principles of Genetics,* 7th edn. McGraw-Hill. 2001.

Sambrook Fritsch and Miniatis. *Molecular Cloning: A Laboratory Manual.* Cold Spring Harbor Laboratory Press. 1995.

Snustad, P.D., Simmons, J.M. and Jenkins, B.J. *Principles of Genetics.* John Wiley & Sons Inc. New York. 1997.

Strachan Tom and Read Andrew, P. *Human Molecular Genetics,* 2nd edn. Taylor & Francis. London. 1999.

Weaver. R.F. and Hedrick, P.W. *Genetics.* Wm. C. Brown Publishers. Oxford, England. 1995.

ARTICLES

Robert Bud. "Molecular Biology and the long-term History of Biotechnology." In Arnold Thackray (ed.). *Private Science: Biotechnology and the Rise of the Molecular Sciences.* University of Pennsylvania Press. pp. 3–19. 1998.

Philip Pauly. "Modernist Practice in American Biology." In Dorothy Ross (ed.). *Modernist Impulses in the Human Sciences.* 1870–1930. Johns Hopkins University Press. pp. 272–289. 1994.

Jacques Loeb. *The Mechanistic Conception of Life.* University of Chicago Press. 1912.

Wells, H.G. "The Limits of Individual Plasticity." *Saturday Review.* 79:89–90. 1895.

Paul Rabinow. "Towards Biotechnology." *In Making PCR: A Story of Biotechnology.* Chicago University Press. pp. 19–45. 1996.

Hans-Jörg Rheinberger. "Beyond nature and culture: Modes of reasoning in the age of molecular biology and medicine." In Margaret Lock, Alan Young and Alberto Cambrosio (eds.). *Living and Working with the New Medical Technologies.* Cambridge University Press. pp. 19–30. 2000.

Gibbs, N. "The Secret of Life." *Time.* February 17, 2003.

Wade, N. "DNA the Keeper of Life's Secrets Starts to Talk." *The New York Times.* February 25, 2003.

Karow, J. "Reading the Book of Life." *Scientific American.* February 12, 2001.

INTERESTING LINKS

http://www.ncbi.nlm.nih.ogv/entrez

http://www.biology-online.org/tutorials/home.htm

http://www.biozone.co.uk/index.html

http://www.accessexcellence.org/RC/AB/BC/

http://www.cato.com/biotech/bio-info.html

http://www.dbt.org

INDEX

www.ingramcontent.com/pod-product-compliance
Lightning Source LLC
Chambersburg PA
CBHW031805190326
41518CB00006B/200